FOR LOVE OF LAKES

FOR

LOVE OF LAKES

DARBY NELSON

MICHIGAN STATE UNIVERSITY PRESS | *East Lansing*

⊖ The paper used in this publication meets the minimum requirements
of ANSI/NISO Z39.48-1992 (R 1997) (Permanence of Paper).

Michigan State University Press
East Lansing, Michigan 48823-5245

Printed and bound in the United States of America.

18 17 16 15 14 13 12   1 2 3 4 5 6 7 8 9 10

LIBRARY OF CONGRESS CATALOGING-IN-PUBLICATION DATA
Nelson, Darby.
For love of lakes / Darby Nelson.
p. cm.
Includes bibliographical references.
ISBN 978-1-61186-021-4 (pbk. : alk. paper) 1. Lakes—United States.
2. Lake ecology—United States. 3. Water—Pollution—United States.
4. Human ecology—United States. I. Title.
QH98.N46 2011
551.48'2—dc22   2011008835

THE DAVE DEMPSEY ENVIRONMENTAL SERIES

Book and cover design by Charlie Sharp, Sharp Des!gns, Lansing, MI
Illustrations produced by Katherine Darnell, www.kldarnell.com

g green
press
INITIATIVE

Michigan State University Press is a member of the Green Press Initiative
and is committed to developing and encouraging ecologically responsible publishing practices. For more information about the Green Press
Initiative and the use of recycled paper in book publishing, please visit
www.greenpressinitiative.org.

Visit Michigan State University Press on the World Wide Web
www.msupress.msu.edu

For Geri,

    my paddling companion for life.

Landscape is our unwitting biography, reflecting our

tastes, our values, our aspirations, even our fears.

If we want to understand ourselves, we would do

well to take a searching look at landscapes.

PIERCE LEWIS

CONTENTS

ACKNOWLEDGMENTS

Many people have instructed, inspired, and encouraged me in the writing of this book. First and foremost I thank my wife, Geri, my first reader, critic, and adventurous companion. Her steadfast support, undying encouragement, and endless assistance and patience made this book possible. My mother, Margaret, first introduced me to lakes as a toddler and loved lakes more deeply than any other person I have known. She planted the seed that became this book, though she did not live to see it grow.

I am particularly indebted to those farsighted people who had the wisdom to establish the Loft Literary Center in Minneapolis, providing a place for novice writers to develop their skills. Many thanks to outstanding Loft instructors Cheri Register, Mary Carroll Moore, and, most particularly, Elizabeth Jarrett Andrew, for their advice, instruction, and encouragement.

I am deeply indebted to dear friends Jim and Marlys Terrian, who read each of the draft chapters, often several times, challenged assumptions that helped me clarify my arguments, and encouraged and supported me throughout. I am also grateful to fellow writers Judy Helgen, Tom Anderson, Douglas Owens-Pike, and Val Cunningham for their invaluable feedback, a healthy mix of tough criticism and wholehearted support. I also thank Sue Leaf and Laurie Allmann for their input at the outset of my journey with the book. Thanks also to John Helland and Barb Coffin for their review and suggestions on the full manuscript in one of its later revisions.

Deep thanks to David Dempsey for his enthusiastic support and encouragement and for interesting a publisher to take a close look at the manuscript. His guidance and confidence played an important role in seeing a manuscript become a book. I also thank Julie Loehr at Michigan State University Press for her enthusiasm for the manuscript from the outset and her ongoing help.

An early version of the "Shield Lakes Icon" was initially published in the *Boundary Waters Journal*. I thank Stuart Ostoff, publisher of the journal, for permission to use a modified version here. I also thank Pat Allen and Andrew Hall of Lincoln, Massachusetts, for the wonderful insights they provided about Thoreau and Flint's Pond, and their deep commitment to protect that lake from harm. Those discussions changed the trajectory of the book. Thanks also to the wonderful librarians at the Concord, Massachusetts Public Library for directing me to key references.

Finally, I must thank the following people for their wisdom and inspiration: Henry David Thoreau, Aldo Leopold, John Muir, Sigurd F. Olson, Rachel Carson, Sam Eddy, Jim Underhill, Jerry Cole, Willard Munger, Myron (Bud) Heinselman, Scott Russell Sanders, and Herb Brooks.

Introduction

We are a landscape of all we have seen.

ISAMU NAGUCHI

My most vivid childhood recollection of lake came the day Mother rented a rowboat at the municipal dock in the town where she was born, loaded my sister and me, oars and lunch, bathing suits and towels, and rowed us up a long stretch of shore to a beach at the city's park. Mother was not an expert with oars, and my sister and I were too young to help with the rowing.

We made slow, zigzagging progress over the placid water—and that made all the difference. I leaned into the side of the wooden boat and peered overboard into sharply clear water, and lake filled the sensory slate of my young mind. I had names for few of the fascinating things I saw. A mysterious forest of skinny green stems with broad wrinkled leaves reached up out of the yellow-brown depths toward the boat. Tiny fish darted to hiding places within the greenery. Spotted leeches, longer than my hand, undulated in purposeful fashion through the green forest. A tiny red globe of an insectlike creature rested on the tip of a submerged leaf. But, above all, I remember the crystal clarity of the water.

Much time passed before I returned to this lake of my childhood. A murky essence has replaced the lake's once vibrant clarity. Other lakes of those growing-up years also show deterioration. I am not alone in my disappointment. Others tell me of standing in water up to their waists in their own lakes of youth, and of seeing the whites of their toes. Now they cannot see deeper than mid-thigh. I hear of beaches closed, fish consumption advisories in over thirty states, and lakes becoming pea soup or choked with impenetrable masses of aquatic plants. Why? How?

We say we "love" our lakes, and crowded shores and the crush to buy lakeshore at astronomical prices suggests we speak truth. Yet our lakes deteriorate, and much of the deterioration results from our own actions. In healthy relationships we care for and protect what we value. What explains the paradox?

Starting with Thoreau in the 1840s, scientific understanding of lake systems has advanced in increments for over 150 years in America. A series of researchers in Illinois, Wisconsin, Minnesota, and New England filled in many of the gaps in lake ecology, giving us a general framework of understanding. Although more details are yet to be discovered, we understand the basic framework of lake ecology well enough to avoid their degradation. Lack of scientific knowledge does not explain the paradox, though lack of understanding of that knowledge by the public might.

Might human perception play a role in the paradox? We can see and hear and touch only a lake's surface and shallow shores. Most of the lake lies hidden from our senses. We are a highly visual species. Can we form accurate perceptions of these bodies of water that we can see so little of? Behavior is the child of perception. Perhaps the paradox is entangled with how we perceive lakes, with how we form understandings.

Thoreau, with his acute power of observation and tendency to record much of what he saw, made a significant early contribution to the emerging science of terrestrial ecology when he worked out the concept of forest succession, notwithstanding his lack of formal training in science. Ironically, despite his intense love of lakes and the considerable amount of time he spent observing and writing about them, to the point of being called by some ecologists America's first limnologist, Thoreau never got beyond a very rudimentary understanding of lake ecology. Was it because he could not see lakes in the same sense he could see the forest, could not "see" what he could not see?

Our choice of metaphor may also influence how and what we perceive. We talk of time as the river flowing. I never questioned the implications of that metaphor until I was struck by the words of Professor Dave Edmunds, Native American, on a display in the Indian-Western Art Museum in Indianapolis. Edmunds wrote, "Time as a river is a more Euro-American concept of time, with each event happening and passing on like a river flows downstream. Time as a pond is a more Native American concept of time, with everything happening on the same surface, in the same area—and each event is a ripple on the surface."

If I think of time as a river, I predispose myself to think linearly, to see events as unconnected, where a tree branch falling into the river at noon is swept away by current to remain eternally separated in time and space from the butterfly that falls in an hour later and thrashes about seeking floating refuge.

But if I think of time as a lake, I see ripples set in motion by one event touching an entire shore and then, when reflected back toward the middle, meeting ripples from other events, each changing the other in their passing. I think of connectedness, of relationships, and interacting events that matter greatly to lakes.

My puzzlement over the deterioration of lakes, despite our love for them, finally bubbled over. I resolved to undertake a journey of exploration to investigate the relationships between people and lakes. I also set out to get a glimpse of our lakes' future.

This book is my account of that journey. It is a cut-off blue jeans and soggy tennis shoes journey, a journey of paddling and wading, listening and sniffing, turning over stones and touching, and reflecting and examining the birthing chamber of perceptions. A journey that takes us to large lakes and small, lakes in my home landscapes of Minnesota and Wisconsin, but also to Canada, Illinois, New England, and, ultimately, Walden Pond.

This journey must begin for me on Rainy Lake, a large lake on the Minnesota-Ontario border. My canoe and I have launched and tasted lake water together uncountable times, and each time the exhilaration is unbounded. How glorious to be *under way!*

LANDSCAPES

Lake Magic

What we call landscape is a stretch of earth overlaid with memory,
expectation, and thought. . . . Landscape is what we allow in
through the doors of perception.
SCOTT RUSSELL SANDERS

I stand again on the shore of Rainy Lake. Its glassy surface shimmers before me. Islands, mounds of green, ships becalmed, dot the placid surface into the distance. The rose-colored mirror turns peach then rich golden-yellow in the rising morning sun. All is silent. Earth rests.

Deep feelings of joy, of belonging, envelop me. Boundaries melt. I seem as one with water, rock, and lily, all part of a magnificent whole.

While such feelings have arisen in me at other times in other places, all have occurred in the presence of water and most frequently, as now, by the side of a lake.

I have not visited these waters since my college days. I have returned to this lake to renew my relationship with a beautiful body of water that awakened in me a deepened interest in the natural world. My personal lake journey began here. I seek sharpened senses to see with fresh eyes this world of lake.

I load my gear into my canoe, launch and paddle slowly from the dock, nose around a small bedrock point, and edge out onto a broad sheet of water, a pane of glass extending for miles, where not a ripple disturbs the reflective surface, where water becomes sky and sky becomes water in seamless union.

I peer through the mirror into the depths. The water is clear but tinted brown, the color of water that seeps from a bog. The canoe passes slowly over a Lilliputian forest of pondweeds. Stalks bearing clusters of skinny, grasslike leaves disappear into the depths. The canoe glides effortlessly through the placid water as I round a point and turn east. A houseboat throttles along ahead of me towing two small fishing boats.

How delightful to be back on the lake. My first visit to Rainy long ago occurred at this same place. Our family was intimately familiar with lakes nestled in outwash sands near the Mississippi headwaters. Rainy Lake was my first exposure to the bedrock lakes of the ancient rock core of North America, the Precambrian Shield. Rounded bedrock domes, humpback islands, hidden coves, and open horizons unleashed wonder and the spirit of adventure.

. . .

This trip is unusual in one important respect. My wife, Geri, is not with me. Also a biologist, she loves paddling and camping adventures almost as much as I do. She would be here, but her teaching schedule ruled it out this time. More patient than I, she believes in checklists. I would not have forgotten my tennis shoes back at the car had she been along.

Miles pass. I am making good progress, but a stand of red pine in a cove invites me to stop for the day. Prudence calls for easing muscles slowly back into their paddling routine the first day of the first outing of the year, so I accept the pines' invitation and pull to shore. I slip comfortably into camp routine. Packs unloaded and tent pitched, I crawl into the tent, unroll my sleeping bag, and glance out the rear window. A red fox limps past toward the lake. He returns my stare.

Though not designated an official National Park Service campsite, a blackened fire ring, hard-packed ground, and the absence of dry wood nearby reveal many have camped here before me. The search for firewood gives me an excuse to explore. The camp sits among pine growing on a small hill of gravelly sand, a rare break in the ever-present gray schist bedrock, and undoubtedly the source of the sand on the beach. A seam of milky white quartz slices through the gray stone at the south edge of the grove. Sometimes associated with gold deposits, quartz seams fueled a short-lived gold rush here over a century ago. A foot-wide stripe of pale yellow pollen forms a bathtub ring along part of the shore. A white-throated sparrow and a hermit thrush sing a duet to close the day.

The gimpy fox returns soon after I open packs to fix breakfast. As I step back to the tent he hobbles, as fast as a fox with three working legs can go, directly to a pack, grabs a plastic bag in his teeth, and scampers off onto a low rock ridge. Luckily for my menu he got the bag with my *Field Guide to the Birds*. It falls to the ground a few steps away. He abandons the empty bag and stands at the base of the ridge, watching.

As I break camp and load the canoe I remember the piece of salmon skin, waste from the salmon sandwiches I'd brought from home for supper last night. I hold it up for the fox to see, set it on a rock, and step away. He limps to it and gulps it ravenously.

A Twin Otter airplane, kayak slung below, flies overhead as I launch my canoe. A labyrinth of islands confronts me. I stop paddling and locate myself on the map, an act reminiscent of my first explorations of Rainy in my college days.

The previous fall my parents had moved to International Falls, on the west end of this large lake, only months after I had left home for the university. I had returned to my family for the summer and to a job in the town's paper mill, having changed my major every few months that freshman year. Actuary? Geologist? Historian? Teacher?

My motorboat adventures on Rainy Lake occupied my days off from work that first summer and the next. I traveled countless miles into bays, around points and between islands. I discovered a long abandoned gold mine and whaleback slopes of smooth bedrock rising from the depths to form land. I also discovered the open horizon of the Brule Narrows.

On one trip past the Narrows to the far eastern end of the lake, I discovered Sand Bay Island. Beaches of sand are uncommon on the bedrock lakes of northeastern Minnesota and adjacent Canada. To stumble on such a beach is a delight. This island had three! Later that summer I came back to the island with two friends for an overnight camp. In my excitement to return, I forgot to bring a sleeping bag and spent the night curled beneath a stiff canvas tarp. I slept not at all, anticipating the next day's explorations. The lake had become enchantress.

I find myself on the map and guide the canoe into a maze of islands. Miles away lies Soldier Point, a long skinny peninsula that forms the American side of the Brule Narrows, my destination. But on the way I must first visit an old stony friend. I pass through a labyrinth of islands and now see the large whitish boulder in the distance that marks the entrance to Cranberry Bay. I first met this massive stone in my college days when it served as my faithful landmark.

There it waits, unchanged. I paddle toward it to renew our relationship. It sits in several feet of water like an enlarged VW beetle. Its domed top projects seven or eight feet above the level of the lake. Uniformly light colored and granitic, it bears no resemblance to the dark gray schist of nearby points and islands. Geologists call this a glacial erratic, a large rock carried by moving ice away from its point of origin to be abandoned, like an orphan, somewhere else. A four foot mass of lichens paints a deep ochre color on the stone's north face, giving the light-gray rock a touch of pizzazz.

So many other lakes have lost their youthful clarity and vigor since I last was here. I rest my hand on the rock. A reassuring sense of groundedness returns. Here I know where I am.

I eat lunch on a small nearby island and resume paddling east. In the space of half a mile I notice three basketball-sized stones with the same composition as my landmark rock. As I break out of islands, I see what appears to be the same rock type forming much of the Canadian shore several miles away. It's as though the stones were Hansel and Gretel crumbs my sentinel rock dropped along its line of travel so it could find its way back to the mother formation.

I plan to camp on one of the National Park Service campsites shown on my map. I approach it but discover it already occupied. Muscles, sore and complaining, must suffer a bit longer. I move slowly along the shore searching for an opening in the undergrowth that can serve as my camp, and find it in a tiny cove deep in Lost Bay. I squeeze the tent between two balsam fir trees. Though the woods are abuzz with mosquitoes, they haven't yet discovered the newly arrived food at the shore.

A pastel salmon sky reflects itself in perfect mirror image on the glassy water. A herring gull lands on the water and approaches shore as if accustomed to getting handouts. I refuse to cooperate. Snorting sounds of flustered deer begin shortly after I crawl into the tent and continue into the night. I have apparently pitched my tent astride their trail to the lake.

. . .

This morning as I break camp, a resplendent pair of bufflehead ducks and two pelicans float silently on water smooth as glass in my cove. The clear sky portends sunburn. I put a large band-aid on my nose in lieu of sunscreen forgotten at home. The paddling is easy. At noon I pull my canoe onto the low bedrock shore of Soldier Point. After lunch of rye crisp, cheese, and a handful of raisins, I explore.

An outcrop of bedrock, bumpy with small protruding garnets, becomes my seat. Eons of weathering have eroded away the softer schist exposing the garnets, among the hardest of minerals. Water four inches deep teems with life beside me. A mass of energized whirligig beetles whirl and gig in random, seemingly nonsensical motion across the water surface. Such energy. Such exuberance. The antidote to pessimism. Sometimes they bump into one another, setting the collided pair buzzing around each other. Two stay together for several swings around a three-inch circle. Maybe it's a mating dance.

A two-inch-long fish rests on the rocky bottom as if he's asleep. Dark speckles on his back and the W-shaped splotches along his sides identify him as a Johnny darter. It's been years since I've seen one. The diminutive darters are relatives of perch and walleye. They dart almost faster than the eye can follow. Male Johnnies' parental instincts are particularly endearing. They not only clear a spawning chamber, often beneath an overhanging rock, but hover over the eggs, fanning and protecting their brood. My Johnny works his way along a thin brown film coating the rock, feeding I suspect, and now disappears into a crack.

As I reach into the water to move a stone, myriad bits of life streak for cover—except for an inch-long object moving slowly across the bottom. It is composed of two elongate slivers of wood, several small twigs, a flat green piece of plant stem, and some stem sheath, all glued together—the accoutrements of a bag lady.

Larval caddisfly peeking out of protective case.

I pluck the object from the rock and lift it into the air. In moments a brownish snout with two tiny eyes peers out from one end, as though puzzling about what is happening to the neighborhood. Finally she fully emerges onto my finger, revealing three pair of legs immediately behind her head and a series of body segments behind the legs. This skinny creature is a larval caddisfly, an insect that spends its early life under water to eventually emerge as a flying adult resembling a small moth. I am distressing her. She is not adapted to live in air. I don't mean to tease her, and I lower my finger into the water. She retreats into her case as it settles to the bottom.

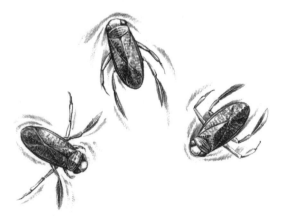

Some water boatmen lay eggs on crayfish for protection.

Waters of Saginaw Bay caress the bedrock across the narrow point. Protected from the northerly breeze by the point, the pool here is glass smooth. Creatures the size and shape of tiny safety pins row themselves jerkily through the water with oarlike legs. Their name, fittingly enough, is water boatmen.

Many feed underwater on the organic layer covering the rocks. Some rise to the water surface. All share a beautifully symmetric pattern of gold and black lines on their backs. A flat sloping rock seems the boatmen's mating place. Several pair up. The top bug clings to the one below with both legs and mouthparts normally used to gather food. The bottom bug, however, keeps right on feeding throughout the mating. Priorities are clear. Many lone individuals approach amorous couples from behind and try to climb aboard to form a trireme. These intruders soon discover their mistake and quickly leave, with apology I presume.

Open water extends so far east from this pool and the Narrows that the land simply disappears. I remember looking out on this place for the first time. Stretching forever, the open horizon seemed filled with infinite possibilities in a future waiting to unfold. Even now it quickens my imagination.

.   .   .

My fascination with the lake grew on those youthful jaunts on Rainy Lake, leading ultimately into a redirection of my life's arc. Back at the university that fall I found myself in a biology class looking at the silvered letters on the green cover of my course book—*Taxonomic Keys to the Animals of the North-Central States*—and wondering. Me, sitting in a biology class, of all things. In high school I disliked, no, detested biology. Pickled frogs. Pickled perch. Endless terminology. And here I was again. Things were still pickled, pickled, pickled. Dead. Dead. Dead.

One day our professor changed our routine and took the class on a field trip to a lake some miles out of town. Students gathered in the tall grass next to a narrow sand beach that sloped to the water's edge. "Wade into the water with a net and see what you can catch," Professor Eddy called out.

Gentle waves slipped through thinly scattered rushes stirring sand grains in the clear water. I took a net with vial attached and waded in. As I moved along I saw small schools of minnows and a few snails but little else.

I swept the net through the water. It was no match for the minnows. They easily darted aside avoiding capture. I swept the net again then stopped to see if I'd caught anything. I grasped the bottle and held it to the sky.

I stared in disbelief at what I saw inside. The bottle was in motion! Alive with motion! Teeming with darting, wiggling, jerking, pulsing vital bits. The top part of the net dropped from my hand and I clasped the bottle with both hands and brought it close to my face. It was not an amorphous mass of agitation, like riled dust. I now saw the bits of motion as individuals, as distinguishable forms of life.

Many swam in exaggerated jerky motions, seemingly propelled by a backward snap of two leglike appendages emerging forward out of the head. These legs arose from near a dense black spot that looked for all the world like a cyclops eye. Others, smaller and with long pairs of feelers, swam more slowly, trailing odd sacs on each side.

A red wiggling motion on the jar's bottom caught my eye. I turned the bottle to see it more clearly. A worm, red as blood, flailed its long body. Its agitations aroused a much larger creature at its side with large oogly eyes. It was brown-green, the color of the lake bottom. Pairs of legs sprouted from behind its head. I then noticed small seeds on the bottom that suddenly lurched into motion, moved by some invisible force across the jar, then, as suddenly, stopped, resuming their disguise as seeds. Everywhere crowded, random, crazy motion. Everywhere vibrancy! Everywhere! I was stunned.

I slowly lowered the bottle and gazed out over the lake. How could this have happened? I had lived my entire twenty years growing up among lakes. I had played along their shores as a toddler, collected shells of snails and clams and crayfish, learned to swim in lakes, fished in them, camped beside them—not rarely, not sporadically, but often. How could I not have seen, not have noticed these creatures. It took no microscope to see this new world emerging before me. I was dumbfounded and ashamed. I had no explanations. The world suddenly seemed much less certain, much more intricate, and vastly more fascinating. What else had I not noticed? What else around me was not as it seemed?

I looked back into the bottle. The jerkers and squigglers now seemed as messengers from an invisible part of the cosmos, emissaries to my blind, unobservant self. My universe would never be the same. I changed my major to biology and ultimately completed graduate research on the ecology of lake herring in the headwater lake of the Mississippi River. Life became lakes and lakes became life.

. . .

I will camp tonight on Soldier's Point for the first time. Professor Eddy understood the magic of lake when he sent us into the water to see what we could find. That magic has returned. I squeeze the tent between a white pine and a balsam fir. Their lower branches rest on the tent's rain fly. The aroma is intoxicating. I snuggle into my sleeping bag and listen. Waves and land converse, giving voice to the shore. Tonight's conversation is of deep things, of assurance and connection, of home.

Limnos I—Walden Pond, Massachusetts

> *A lake is the landscape's most beautiful and expressive feature. It is earth's eye; looking into which the beholder measures the depth of his own nature.*
>
> HENRY DAVID THOREAU, 1854

I first learned of Henry David Thoreau in high school English class. He seemed an odd sort, squirreling himself away in a tiny cabin by a lake, refusing to pay taxes to support a war against Mexico he opposed, an act of civil disobedience that cost him a night in jail, a writer who refused to hold a steady job, a social misfit who loved nature, particularly that small lake, Walden Pond.

I rediscovered him while searching for a pithy quote to use in a talk I was preparing decades ago on the lakes of Minnesota's Boundary Waters Canoe Area. I found my quote; it tops this page. I also discovered a Thoreau I never knew, an endearing bundle of contradiction, and our first and foremost lake watcher.

He describes his famous pond as "a gem of the first water which Concord wears in her coronet. . . . It is like molten glass cooled but not congealed, and the few motes in it are pure and beautiful like the imperfections of glass. . . . Not a fish can leap or an insect fall on the pond but it is thus reported in circling dimples, in lines of beauty, as it were the constant upwelling of its fountain, the gentle pulsing of its life, the heaving of its breast." He sees the lake through the lens of emotion expressed in the language of the poet.

Thoreau, despite his romantic expressions and protestations to the contrary, also saw lake through the window of reason, of science. He is not usually thought of as a man of science. Academic credentials not-withstanding, that is oversight. Intensely curious and an acute observer, he had the mind-set of science. Particularly in his later years, Thoreau tended toward increasingly objective observations of local natural history. In addition to significant original writings on forest succession, dispersal of seeds, and other aspects of terrestrial ecology, Thoreau also developed insights into limnology, the study of lakes. (*Limnos* means "lake" in Greek.) *Walden* contains some of the earliest objective observations on the nature of lakes.

When town folk concluded Walden to be bottomless, Thoreau rejected such nonsense, tied a rock to a rope, and discovered its maximum depth to be 102 feet. A part-time surveyor, Thoreau calculated Walden's surface area at 61.7 acres. Limnologist Edward Deevey, in 1939, with more sophisticated equipment, measured 61.3. Thoreau discovered that lake temperatures change with depth. He linked the increase in water temperature that occurs over sand bars beneath ice to sunlight reflecting off the bottom. He was the first to record the phenomenon of streaks of foam formed on a lake by a strong wind, now known as Langmuir spirals.

Thoreau noted the differences between lakes rich in nutrients and those nutrient-poor, known today as eutrophic and oligotrophic lakes, respectively. His descriptions helped my students visualize these two principle lake types. Coupled with his many observations of fish, aquatic plants, and other creatures in Walden and other lakes, Thoreau's scientific accomplishments are impressive. Deevey wrote, "It has been shown that Thoreau's curiosity was unusually fruitful when directed toward lakes . . . so that the Concord individualist may with justice be called America's first limnologist."

Thoreau's observations, however, like those of credentialed natural his-tory professors of his day, were purely descriptive. Thoreau, like the others, saw lake as a body of water and a catalog of individual living things, not as functional relationships.

. . .

Ironically, as a self-professed mystic, transcendentalist, and natural philosopher, Thoreau railed against the growing insistence of science in his day that the cold hand of objectivity, devoid of human feeling, was the exclusive path to truth. "The inhumanity of science concerns me, as when I am tempted to kill a rare snake that I may ascertain its species," he wrote. Nature, according to Thoreau, was not detached from self. "The point of interest is somewhere *between* me and them," he wrote.

My training in science demands fealty to the standards of reason as *the* legitimate path to the hard bottom of truth. Instinctively, I distrust the claptrap of emotion. My mind sees the pitfalls of romanticism, of blindly following innate inclinations, in the mucous trails left by slugs on a bike trail a few blocks from my house. Why so many of these wayward mollusks forsake the vegetated sanctuary on one side of the trail to cross tar to the other side, I cannot say. It must have made instinctive sense to the molluscan brain. But their slime trails left glistening in the afternoon sun record the folly of their "truth." The trails start out headed more or less toward the other side, but soon the silvered ribbons begin the twists and turns and double-backs of creatures that have lost their way. Most end in a squiggle that marks their final anguished moments before death by dehydration overtakes them, leaving carcasses dried and fried on the tar. Beware instincts, my mind says, beware what merely *feels* true.

And yet, my science cannot explain the magic pull of the lake, the intense abiding joy I feel on quiet water, cannot explain the power of Thoreau to penetrate my being so deeply.

. . .

Thoreau had a foot firmly planted in each of the ways of "knowing," the feeling-based instincts of the romantic and the measured, numbered, peer-reviewed world that is science. As he grew older his perspectives evolved and his writing took on a progressively more objective hue, even as science evolved to hold more sophisticated understandings of lake in the 150 years after his death.

Thoreau sensed his progressive slide toward the cold barrens of objectivity in the last ten years of his life but seemed helpless to do anything about it. "I have become sadly scientific," he concluded.

Deep Heart's Core

A few years ago in early April four immense earthmovers arrived at the edge of a farm field two blocks from my house in a Minneapolis suburb. The yellow behemoths arranged themselves in an intimidating line, then fell silent awaiting orders. The next week the machines growled into action and fell to rearranging dirt, soil that had nurtured corn and soybeans for as long as I can remember. When the machines left three months later, a large irregularly shaped hole remained in the center of the field. It began filling with water. An earth-tone sign sprouted at the corner of the field proclaiming: "Cobblestone Lake Preserve—Executive Lake Community."

Yellow digging machines followed and mounds of earth arose beside holes the size of houses. "For Sale" signs erupted like weeds across the landscape. The lots sold quickly, buyers rushing to pay a premium for lots on the "lake," which in truth is little more than a storm water retention pond. More than nine hundred lakes dot the Minneapolis–St. Paul metropolitan area. More than 12,000 grace the landscape of Minnesota. Not enough. Our appetite is insatiable.

Another housing development, Diamond Lake Woods, has recently taken root in a farmer's field off the southwest corner of a lake two miles from my house as the cormorant flies. Though no lakeshore is involved, those lots that merely have a *view* of the lake cost half again more than other lots not so blessed.

My time spent stuck in traffic jams with people heading to lake country from Minnesota to Wisconsin, Michigan, Ontario, and New England convinces me that we humans have, if not a flat-out love for lakes, a particularly strong attraction to them.

That attraction has deep roots. In 1910 the popularity of Lake Winnipeg beaches busied a dozen trains a day moving more than 40,000 people between Winnipeg and the lake on holiday weekends. By the early 1900s, housing developers in places like Lakewood, Ohio, had discovered that lakefront lots appealed to homebuyers more than lots in other locations, and by the 1880s the people of Cleveland had begun building summer homes on the shore of Lake Erie. The hundred rooms of the Mineola Hotel drew vacationers to the shores of Fox Lake in northeastern Illinois in 1884. Sixteen years later a rail line from Chicago opened floodgates, pouring hordes more visitors into the Fox River Basin lakes. Decades earlier, in the 1840s, Henry David Thoreau was sharing with readers his infatuation with Walden Pond. I suspect an emotional pull, not conscious reason, determined Thoreau's choice of a lake as his central metaphor.

By the late 1700s landscape artists and poets had discovered England's Lake District. William Wordsworth and Samuel Coleridge, smitten by the landscape, went to live among the lakes and gained fame and following as the Lake Poets. Stimulated by the writings and paintings of a coterie of artists and authors, the number of visitors to the lakes became a flood as quickly as road improvements allowed.

The power of lake's pull is evident even in the early seventh century. Celtic Christians believed resurrection occurred at the site of burial and so took pains to place churches and cemeteries in what were judged especially aesthetic settings, often by lakes, to please the soul for eternity. Saint Kevin chose to live in the Glen of Two Lakes, Glendalough, refusing suggestions by an attending angel that the setting be modified.

Why are we drawn to still waters? The feeling is so natural the question itself appears silly, like asking why we love babies, mourn our dead, or laugh at one circumstance and cry at another. Attraction to lakes is simply another of countless human tendencies. Zoologist Gordon Orians suggests, however, that, "As anthropologists have long known, the values of society that nobody questions reveal most about a culture."

Explaining human behavior has never been easy. Do we go to lakes simply to play? ("Laking" is the dialect word for playing in England's Lake District.) Increasingly, my observations don't support that explanation.

Geri and I once traveled to a state park with our canoe for a lazy weekend at a lake we had never explored. Pleasantly warm temperature, a bright sunny sky, and calm winds made the lake irresistible. Cabins and homes lined the waterfront. It seemed a wonderful opportunity to poke along the shore by canoe and see what people actually do at the lake.

At the first cabin a man threw sticks into the water for two bouncy golden retrievers to fetch. Five houses later a woman mowed her lawn, but there activity ended. Despite the gloriously inviting weather, lawns and docks showed no sign of life. No one swimming, no one wading, no one fussing with boats. It was not that cabin owners were not home. Cars parked beside the buildings revealed otherwise. Only the far-off whine of a Jet Ski and a scattered handful of fishing boats revealed the lakescape to be other than abandoned. More buzzing boats and Jet Skis emerged later in the day, but most of those came from the public landing. Active lake use by cabin owners seemed almost insignificant considering the number of dwellings on the lake.

People flock to Minneapolis lakes, particularly Lake Calhoun, all summer long. Few visitors are drawn there to fish or boat, though some sail, canoe, or kayak. Most come to walk or jog, bike or roller blade or sun themselves. So many people visit the lake that safety concerns forced the Park Department to build two parallel trails around the lake to separate folks on wheels from those on foot. People sunbathe nearly everywhere grass or sand covers the ground. I've observed the same flock-to-the-lake behavior as far from the northern lake country as Town Lake in Austin, Texas. But why do walkers and joggers, bikers, bladers, and sunbathers crowd themselves onto the paths of such places as Lake Calhoun and Town Lake, or any other lake? Nothing inherent in these activities necessarily links them to water.

I live two blocks from a large regional park that contains nineteen miles of excellent bike trails. Trails meander pleasantly across gently rolling terrain through open grassland and wonderfully shaded stretches of woods. But walking, I meet more people in two minutes on the paths around Calhoun than I do in four hours at the park. And no one sunbathes at the park anywhere—except crowded around the shores of the man-made swimming pond. I am unconvinced that we are attracted to lakes purely to play. Rather, I suspect play provides an excuse to go to the lakes.

A Florida study provides insight. A large number of lake users were asked to rank the amount of time spent at each of twelve lake-user activities, from fishing to boating, swimming to Jet-Skiing, and everything in between. The number-one use by far? Just sitting and enjoying the lake.

Might I owe my attraction to lakes to my upbringing? Mother was born a block from a lake and appears to have been imbued with the spirit of lake with her first breath. As early as I can remember she carted my sister and me and any cousins that happened to be in town to lakes. Her photo albums reveal her lake habit continued long after my sister and I left home, on into her eighties, until hints of Alzheimer's began to appear. She never had the least interest in fishing or waterskiing or boating, nor did she want to own a lake place, as though she loved all lakes and refused to settle on just one. While she loved to swim, she spent most of her time basking on shore. Near the end of her life, when dementia had robbed her of reason, memory, understanding, and even the ability to recognize members of her own family, she once asked me, "What time of year is it?"

"Early autumn," I replied, "Summer is almost over."

She paused for a moment, then, with a dreamy smile said, "Oh, haven't we had some wonderful swims this year?" She actually had not swum in a decade. Although Alzheimer's had erased much of her mind, it had failed to dislodge the powerful images of times at a lake. The quiet water had penetrated to the depth of her being.

· · ·

Gordon Orians and others tried to learn if humans are instinctively drawn to certain landscapes. They surveyed a large number of people across a great diversity of ages, cultures, and countries worldwide and discovered that humans do favor a particular landscape, one of short grass, scattered trees, and shrubs set among gently rolling hills by *standing water.* That such a wide variety of people embraced the same image suggests that we may indeed have an innate preference for place.

But might this preference be learned? That explanation would seem more plausible if people's description of their ideal landscape matched the landscape where they lived. It did not. Preference for the savanna-by-the-lake landscape was most intense in children.

We reveal our landscape preferences in our actions. Homes along golf courses, with their strong resemblance to the savanna landscape, command premium prices. We incur the ire of neighbors if we do not keep lawns closely mowed. When a resident of my city let her postage-stamp lawn grow unmowed, allowing milkweed to thrive to benefit monarch butterflies, neighbors complained. Acting on an ordinance prohibiting such "unsightliness," the city mowed her lawn and sent her a bill for the work. The rights of town folk to live in a landscape of the mind superseded the rights of a homeowner's choice. If landscape lacks lakes, and we have money enough, we create our own and build our houses around them.

.   .   .

Some argue that art expresses our deepest needs and feelings. I believe them. Thomas Smith's 1761 engraving, *Derwent Water at Crow Park,* became the most popular image of the English Lake District. Two figures stand in the left foreground under the branch of a large tree. Cows and several clumps of trees balance the picture on the right. A closely cropped pasture slopes down to a lake. Several cows stand in placid water as a person pulls a boat onto a narrow peninsula, the boat casting a shadow on the mirror surface. Close-cropped vegetation, scattered clumps of trees, and peaceful water, it was as though Smith had been hired as illustrator for Orians's research conclusions. That the artist's image illustrated so precisely the findings of anthropologists 230 years later struck me as more than coincidence. Gordon Orians wouldn't have drawn humanity's favorite landscape any differently himself. All is quiet in the picture. Earth and soul rest.

.   .   .

Absolutely nothing about Rainy Lake or Cobblestone Lake would make you think of Africa. Yet our attraction to lakes may have everything to do with Africa. Maps of the eastern side of that continent show many lakes lying along a gently curving crescent, running from the Red Sea in the north some two thousand miles south to Mozambique. This string of lakes marks the location of the magnificent East Africa Rift Valley, with the great Serengeti Plain and its vast herds of zebras, wildebeest, antelope, and the big cats that eat them. Mount Kilimanjaro rises midway along the eastern flank of the broad valley.

This crescent also marks the birthplace of our species.

The Rift Valley contains a treasure trove of fossils of our lineage. Excavations at the fossil digs in Kenya, Ethiopia, and Tanzania have not only produced hominid fossils, but also reveal that these protohumans lived on the shores of large marshy lakes. The earliest fossils of *Homo sapiens* and their stone tools have also been unearthed at former lake shore sites in this great valley of East Africa. We have been drawn to lakes from our very beginnings.

While people affect lakes today, insights gleaned from the Rift Valley suggest that in times past lakes may have affected our ancestors. Two questions lurking in the history of our species may both find explanation through our long ancestral association with lakes.

Here's the context of the first. The fossil record suggests our family tree separated into two main branches three million years ago. Both groups thrived for a considerable time. Then, about a million years ago, the Australopithecine line became extinct while the Homo line continued, ultimately producing us. Extinction—why them and not us?

The most successful Australopithecines lived among mountains, in river valleys, and in closed forest settings. By contrast, Homo fossils are more typically found along ancient lake margins. But habitats can change over time. We think of glacial ages as phenomena of the north, but the ice age affected climate around the world, including East Africa. Though no ice flowed down the Rift Valley, the region's climate became progressively more arid. Forests diminished, rivers dried up, and lake basins shrank.

The deep lake basins of the Rift Valley became the only permanent, reliable sources of water. Nutrition scientist C. Leigh Broadhurst claims,

"In another geological setting, without the numerous Rift basins to retain water, a climate of such aridity would not have permanent lakes." And, one can surmise, perhaps not Homo either. Imagine the terrible times this climate change imposed on the Australopithecines that typically did not live next to lakes. How fortunate were the Homo species for their lake-loving ways. Did their intimate association with lakes spell the difference between survival and extinction for the Homo line? One cannot say for sure. But we do know that only the lineage that lived beside lakes left descendents to read these words.

The second puzzle in the human story surrounds the explosive growth in size of the "thinking" part of the Homo brain which, along with changes in pelvic structure to allow the birth of big-headed babies, gave rise to *Homo sapiens*, "wise man." The speed with which the brain grew was astonishing by geologic time standards, and the precise agent that enabled this remarkable change may be linked to lakes.

Our brains have a far greater fatty composition than other body tissues. The body cannot manufacture several key fatty acids needed for brain tissue. They must be in our food. Broadhurst believes the brain's explosive growth could not have happened without a plentiful supply of these special fatty molecules in the diet, especially two long-named substances known more simply as AA and DHA. Although AA is available in land animal meat and eggs, DHA is not. Broadhurst points out that no such deficiency exists in food obtained from a lake, that the flesh of freshwater fish has a DHA and AA composition that is "closer to that in our brain . . . than any other food source known."

An excellent source of protein as well, fish and shellfish of the Rift Valley lakes offered the perfect diet to supply the high nutritional demands of doubling the size of the ancestral human brain. The fossil record links fish bones, barbed spear points, and fish trapping weirs and dams with early human sites.

Was life at the lakes key to the origin of our intelligence and the creation of our species? Broadhurst concludes, "If hominid diets were consistently deficient in [the AA and DHA molecules] the uniquely complex human neurological system could not have developed. . . . We hypothesize that consistent consumption of fish, crustaceans [and] mollusks . . . from lake margins provided [an easy] means of both initiating and sustaining growth of the cerebral cortex." I don't know if Broadhurst is right, but regardless, it is clear we humans have a longer, deeper relationship with lakes than generally realized.

Another observation raises even stranger questions about relationships between humans and watery places. We are the only member of our close family, the Primates, able to efficiently swim, produce chubby babies, have fat attached to our skin, be nearly hairless, and have conscious control over breathing. These traits we share with only one group—the large marine mammals—whales and dolphins and seals—and no other animals except the near hairless elephants and hippos. Puzzling.

· · ·

I recall a conversation a year ago with my brother-in-law in a fishing boat on a small lake in northern Minnesota. Chuck has lived in a nearby town much of his adult life and has fished the lake for years. Though cabins and homes line parts of its shore, the lake retains a pleasantly natural appearance. I asked Chuck, "What do you suppose most motivates these cabin owners to have a place on the lake—the fishing?"

"No, I don't think that's it," he replied. "I bet half these people never wet a line. Even people who halfway enjoy fishing do it rarely. For most people, I think it's the view. Someone said that for property what matters most is location, location, location. Up here for a lot of people it's the view, the view, the view."

· · ·

I find myself too often visiting hospitals and nursing homes these days, as the frailties of age catch up with my parents' generation. I've become acquainted with the galleries of pictures that always line corridor walls in such places. Most pictures are of natural landscapes and half, usually more, include quiet water. I find it more than coincidental that even here—particularly here—we find the view, the view, the view that touches the deep heart's core. When we turn to these beautiful places we find the landscape of home.

August Epiphany

*Maine has a reputation for beautiful pristine lakes, clear blue
waters, loons calling, and pointed firs framing the shores. But
Maine lakes have been showing signs of declining water quality
over the last 20 years. We are losing the clear water, the loons
and the pointed firs.*

BARB WELCH *and* CHRISTINE SMITH

Driving the road that hugs the lake's south shore, you would notice little to
set Diamond Lake apart from hundreds of others nestled in gently rolling
farmland of the Minneapolis–St. Paul outer suburbs. When I first visited
decades ago, the lake had lost the crystal clarity that led early settlers to
name it Diamond. It has a tendency to "green-up" as its waters warm in
early summer, not unlike many other shallow lakes where a rich nutrient supply stimulates algal growth. Though Diamond attained perhaps a
greener state than others, it was but one of many lakes whose aesthetic
qualities had deteriorated through nutrient enrichment. "Nutrient-rich."
How benign, even positive, the words sound.

My visit to Diamond Lake one late August day shredded that perceptual illusion as surely as the first swim of the season in a lake recently
ice-free convulses the senses. I arrived at the public landing on the lake's
south side at midmorning, my first visit to the lake since the previous May.
A red Geo Metro with assorted nicks and rust spots, door open and radio
blaring, sat at the edge of the gravel pad. A man and his young daughter
sat in aluminum lawn chairs in the water, fishing. After watching me pull a

hidden log out of the weeds to rest my canoe on, the man said, "I see you've been here before."

"Yes," I replied. "I use this driftwood to keep my canoe off the sharp rocks while I load it. How's fishing?"

"We seen some nice ones jumpin' off that brush over there, looked like crappies. Haven't caught any yet," he replied. I explained that I also had seen fish jumping here back in May, but they looked like carp. He wished me luck as I finished putting paddle, life jacket, journal, lunch, and, for the first time here, snorkel and mask into the canoe and pushed off.

Though the canoe floated on water less than a foot deep, I could not see bottom. The lake had never been particularly clear on previous visits, but this time it seemed especially green. A green film, with a ribbed appearance like ripple marks in sand, covered the calm water surface. I recognized the windrows of flourlike material as a kind of bacteria commonly known as blue-green algae. There would be no snorkeling. I wouldn't have seen beyond the end of my nose. I felt the fool. I paddled out from shore to discover the lake beyond was only slightly less green.

Water drops falling from my paddle splashed out transparent circles on the green film, leaving a trail of droplet footprints in the canoe's wake. More blue-green algae appeared, this time in patches along the shore. I pulled a bottle from my mesh bag and scooped up a sample, only to learn that here the green was not a film at all, but a glob, a disgusting green clot an inch thick, like congealed paint. I shook the bottle, breaking the clot into tiny clumps of algae.

The green-tinted water stretched to shores in all directions. The lake appeared as a vast vat of algae, an incubator of cells too small to be seen as individuals but dominating to the eye en masse. Zillions of cells. I had never before seen a lake's blue-greens dense enough to form clots. I admit to grudging respect and intense curiosity about any creature that can so effectively take over a space, disgusting though it may be. More globs of green appeared. Disillusionment grew.

An odd *pucker-smack* sound emerged from a stand of cattails. I glided to a clump of plants and waited. Pucker-smacks punctuated the stillness up and down the cattail fringe. Something wiggled the clump next to my canoe. Then another pucker-smack—close. It sounded like fish feeding at the surface. While I had seen a few fish jumping, the pucker-smacks came from the dense tangle of plants that choked off spaces between cattail clumps, hardly the habitat for fish. I waited, hoping to learn the sound's source. Silence. Some minutes later, no closer to identifying the pucker-smackers, I resumed paddling.

A dense, sickly mat of emaciated plant material confronted me as I entered the next bay. I strained at the paddle, gaining only inches with each pull. I could not identify the plants composing the mat at first. They had lost all but a few of their leaves, leaving long, bare, brown stems and a series of nodes where leaves once attached. I thought of cancer patients whose therapy had cost them their hair. On closer look I finally recognized the material as Elodea, the plant so familiar to my students in lab back at the college. But where were the robust stems and the brilliant green leaves that bubble off oxygen like effervescent fountains? Where was the vibrancy, the life? This was decidedly not the Diamond Lake I visited back in May.

A patch of yellow-green algae brightened a pool within the mat. Then I noticed two strange gray clods embedded in the mat surface. One was over a foot long. I hesitated to touch it. It looked like mold, like something that had grown in a long-forgotten canning jar in a dingy cellar.

A fish swirled beside the canoe, moving the surface of the stewlike mat, creating two circular openings, like a pair of eyes through which I could peer into the unholy mess. A channel opened in the mat, a thinning in the distressful brown. A polygonum plant, with white and pink column of flowers and a single white water lily, tried to spread cheer. Several small damselflies sat motionless within arm's reach on one of the lily pads.

Nowhere did I detect the familiar, inviting smell of lake. Stench pervaded the air. Not the odor of rotting fish, another kind of stink from putrefying plants. The whine of a lone Jet Ski far out in the lake broke the silence as I maneuvered the canoe out of the bay. I pulled my way through more mat with the paddle and came on two fish backs protruding above the mat: carp. One made the pucker-smack sound as I watched. Suspicion confirmed. I touched the fish with my paddle, if a creature living in such thickness can be called a fish, and it swirled away. The green mat rose and wiggled as the invisible creature moved beneath. It seemed inconceivable that fish could live in such thick habitat.

Fishermen, limnologists, and lake lovers rue the day carp entered the continent's waters. Not native to North America, *Cyprinus carpio* was brought from the Old World to the United States in 1872 and released into Minnesota waters in March 1883. The fish reached St. Paul and an excited public by rail, dripping with irony. According to Minnesota Fish Commission reports of the time, the rail car's cargo was considered so precious authorities posted guards to prevent theft of the fish, so prized for food in Eurasia. Able to eat both animal and plant matter, very prolific (up to 2 million eggs from fifteen-pound females), and capable of rapid growth, carp quickly spread. This species stirs up lake bottoms and uproots and destroys vegetation, turning clear water turbid, rendering lake bottoms uninhabitable by native game fishes. Carp are partly responsible for the miserable condition of this lake and many others.

Farther along, two walleyed pike, each the length of my hand, lay dead on the surface of the stew. How did those fish get so far out on the mat? I've never expected walleyes could establish a viable population in this lake. Too shallow and too vulnerable to winter kill. Someone must have tried to stock them.

Beyond the bay a lone white pelican swam slowly out from a massive tree that had fallen into the water. Barkless, several large branches reached skyward festooned with several dozen black cormorants, wings outstretched to dry. Pelicans and cormorants eat fish. Carp must be the perpetual special of the day.

I pulled my way forward and stopped beside a hideous island of scum nearly the length of the canoe. A moat of open water separated the scum from the emaciated mat, as though the surrounding plants had withdrawn themselves to avoid physical contact. Depressing shades of olive and green, brown and gray, and blue and a shot of black lay in swirls and loops and circles. A breeze nudged the canoe into the scum, folding it into coarse crusted wrinkles. My bafflement turned to anguish.

I extracted myself from the dreary mat and returned to open water, mind reeling. What happened to the Diamond Lake I knew three months before and the years before that? How had this lake sunk to such a degraded condition? Why was I so surprised, so disturbed by what I'd seen? Couldn't I, shouldn't I, have seen it coming? I felt betrayed by my own failed

Lowered into the water, the Secchi disc measures water clarity.

perception. Though I had never before visited Diamond Lake in August, intellectually I understood that plant and algal growth commonly peak in mid to late summer. Couldn't I have extrapolated my observations in May into conclusions about August? Had my scientific knowledge prevented full comprehension, holding my mind in some intellectual box, keeping me at arm's length from feeling a deeper sympathetic connection to this lake?

I began paddling slowly toward the landing. The red car was gone, but a motor boater attempted to get two young water-skiers up onto their skis. Again and again the skiers failed to get ski-borne, jerked headfirst into the water at the boat's acceleration. Again and again the boat circled back to the bobbing heads, trailing two ropes from its transom, to try again. Finally all gave up and the two people in the water, incredibly, began swimming through the green soup toward shore.

I can't understand what possessed them to swim, inevitably taking in mouthfuls of the colored water. I can't understand what pleasure the fisherman and his daughter found in their lawn chairs in the soupy green shallows at the landing. Jet Skis? Swimming? Fishing? Here? I cannot imagine what perceptions guided the behavior of these others with whom

I shared the lake. How accepting they seemed of the lake's condition. I decided they must not have known the lake's true condition, as I had not known two hours before. They must not have known of the sights and smells in the small bays.

I have experienced many lakes in my life but, truth be told, most were in wilderness in the far north. On most of those lakes, when thirsty, I simply dipped my cup into clear lake water and chug-a-lugged the refreshing liquid. As I looked at the opaque green water beside my canoe, the mere thought made me nauseous. Had these prior experiences distorted my senses, skewed my perspectives, numbed my sensibilities to realities in my own backyard?

I glanced up. Towering thunderheads covered the sky. I paddled furiously toward the landing and the safety of my car.

• • •

Deeper questions arose. Were Diamond Lake's conditions and its willing users aberrations, or were they representative? Was Diamond Lake alone in its crash from grace?

My search for answers began in mid-July the next summer when Geri and I visited over a dozen lakes in the southern part of the state. I intended no detailed study. I simply wanted a quick look at a sample of lakes to compare to Diamond.

I brought my Secchi disc, an object the size of a dinner plate, with alternating quadrants of black and white. The disc is lowered into a lake by a measured rope until its white sections first disappear. The depth to disappearance serves as the measure of water clarity. Clarity serves as a very general indicator of a lake's health, much as a doctor's measure of pulse, blood pressure, and chest sounds indicates the general health of a person. Low lake clarity usually means something has gone awry.

Extracts from my lake journal record our discoveries in four days of exploration.

CANNON LAKE, SHAGER PARK

Fluffy white cumulus clouds. Warm. Greenish water . . . clarity as low as on Diamond Lake . . . two boats and two Jet-Skiers traveling the lake. Several families are picnicking . . . no one swims at the beach. A windrow of deep blue-green algae lines the beach . . . smells like Diamond Lake, unpleasant.

· · ·

LOWER SAKATAH LAKE, SAKATAH STATE PARK

Strange water . . . a bit less murky than Cannon or Diamond, but filled with narrow green spindles a quarter-inch long. "Spindle" algae are so dense on the north shore the Secchi disc disappears two inches below the surface!

I later learned some blue-green algae species can aggregate to form such spindlelike structures. Looking into the lake was like looking into a cocktail glass with a dense cloud of green particles suspended in the liquid as though concocted by an Irishman as a St. Patrick's Day novelty drink.

· · ·

UPPER SAKATAH LAKE, SAKATAH STATE PARK

Spindle algae and swirls of blue-green color mixed with swirls of white, like objects of art, are scattered in small bay. I expect the white is a water mold. Geri notices the water is layered. Greenish surface film on top, then 5 to 6 inches thick of the same green spindles as on the lower lake. The layer of algae below is so thick it looks like a false lake bottom. Two very large pontoon boats, two kayakers, Jet Skis and runabouts ply the lake.

We were paddling the lake to learn of its condition. Why would those others come to it for recreation? A house sat back from the shore maybe a hundred feet, on land no more than two feet above lake level. A bright green grass-covered septic mound sat even closer to the lake. Algal density varied in different parts of the lake. Several dense schools of tiny black fish wiggled as a solid mass away from the canoe. I scooped one fish up in a jar. Baby bullheads, or maybe madtoms. Two dozen cormorants, black as ink, perched silently on the branches of a tall tree, like vultures . . . waiting.

Paddling back to the landing on Lower Sakatah we watch as a pontoon boat serves as the staging area for another boat to get novices up on water skis. Murky water not much different than on Diamond is acceptable to them.

· · ·

LILY LAKE

What delightful relief. Dense stands of white lilies, two kinds of yellow pond lilies, and many other aquatic plants all vibrant green. Secchi depth over 8½ feet! Swallows and dragonflies skim the water surface feeding. What a lovely surprise.

Lily lay in a steeply sloped valley. Not far back from the hill crests, the ground sloped away from the lake. Rainwater and snowmelt as runoff would mostly drain in the opposite direction from Lily.

We visited fourteen lakes in those four days, including immense Swan Lake, once, but no longer, one of the premier waterfowl lakes on the continent, and Bear Lake near the Iowa line. Based on water clarity, smell, and general level of aesthetic appeal, my journal notes ranked Cannon, Upper Sakatah, Lower Sakatah, Volney, Pickerel, and Bear on a par with Diamond. Diamond Lake was no aberration. Degradation had savaged all these lakes to a degree I had not imagined.

No one advocates letting lakes slip into the condition of a Diamond or a Sakatah. How could we have come to this? Our relationship with lakes at that moment struck me as having passed from paradox to pathology.

My thoughts turned to Thoreau's words. None of these lakes could honestly be called "the landscape's most beautiful . . . feature." Though all could be called an "expressive" feature, expressive of what is the question. One survey found that people who live among degraded lakes are more accepting of dirty water than people accustomed to cleaner ones. I don't know whether to cheer or to cry.

Agassiz's Gift

The real voyage of discovery consists not in seeking new landscapes, but in having new eyes.

MARCEL PROUST

Five hours after leaving home I turn off the highway onto a county road and drive for one mile to where the road tees. I pull the car onto the shoulder and open my map to get my bearings. I am searching for a ghost lake, a lake that no longer exists.

Perhaps, in some subconscious way, my childhood fascination with a storybook tale of a boy who swallowed the sea, so his people could walk unhindered over the sea floor to gather fish for food, explain why I am here. How exciting, I thought, to walk the bottom of a lake. What strange new things one might see.

. . .

A great valley several miles wide cuts a broad swath through the corn and soybean country of south-central Minnesota, slashing south then east from the state's western border for 300 miles. Its line runs remarkably straight, as though whatever created it had serious intentions. My family moved to a small town on the floor of that trough when I was six. On move-in day, as my parents carried our few belongings into the north side of the white duplex that was to be our home, I became fascinated by the bluff line of huge hills that dominated the landscape along the north edge of town.

Steep ravines separated the hills. A gravel road climbed one ravine to flat farmland above. Morton Creek arose out of another, its waters eventually reaching the chocolate flow of the Minnesota River at the edge of town. The Minnesota at Morton is not a large river. I once watched a man wade across in hip boots. The river wanders aimlessly through farmland in a ludicrously oversized valley, as though it is lost.

The town sits at the edge of a bedrock platform protruding above the valley floor. A knoll of bedrock extends the school playground into a rolling terrain of rounded rock knobs. I did not understand then, nor when we moved away seven years later, that the immense valley in which we lived spoke of momentous events of the past. I could not know then that, but for a gigantic lake farther up the great valley, there would have been neither exposed bedrock nor steep hills on which to scramble.

Lake Agassiz, named for a bright Swiss natural history professor, extended north from the head of this valley at the Minnesota–South Dakota border nearly a thousand miles to northern Saskatchewan and six hundred miles northeast into Ontario. Although its shoreline changed over time, Lake Agassiz is the largest lake to have existed in North America. This great lake, in its time, covered an area four times the size of present-day Lake Superior, the largest existing freshwater lake in the world. Lake Agassiz is no more, though large lakes in Manitoba and northern Minnesota occupying its deepest depressions remain as reminders of the great lake's expansive past.

.   .   .

My second association with the ghost lake came when we later moved out onto Agassiz's former bottom in northwestern Minnesota, where, as Laurie Allman has written, "The landscape is level enough that caterpillars crawling across the road stand out as topographic relief." I must have heard of the lake by then, but sense it, feel it, believe it? How can you see lake when your eyes tell you of towns and fields of corn, sugar beets, flax, and sunflowers?

.   .   .

I have come to discover the great lake for the first time, to establish a perceptual relationship with a ghost. I cannot touch its water, nor watch its whitecaps roll, nor hear them smash against the shore. I cannot sniff its smell and I cannot taste its sweet waters. Can one experience such a lake?

Louis Agassiz, the lake's namesake, once tried to establish such a relationship with a ghost. His ghost was not a lake but a thick sheet of ice. On July 24, 1837, this brash up-and-coming scientist jolted the scientific world. He set aside his planned address on fossil fishes and laid before the gathered naturalists of the Swiss Natural History Society an audacious idea. Thick sheets of ice of continental proportions had once moved across the land, he said. Glaciers had sculpted the landscape. Nonsense, chorused fellow scientists as word of his claim spread.

William Buckland, prominent English geologist, had launched a full-scale investigation early in the 1800s to explain how landscapes came to be. At the outset, like most scientists of his day, Buckland had no doubt that the catastrophic biblical flood had been responsible. The task before geologists, as he saw it, was to fill in the details, "to confirm the evidence of natural religion: and to show that the facts developed by it are consistent with the accounts of creation and deluge recorded in the Mosaic writings."

But look at the evidence, said Agassiz, look at the evidence. Immense boulders located miles from parent rock formations. Deep parallel scratch marks in bedrock in mountain valleys and elsewhere. Others finding similar puzzling features across Europe. At first Buckland held fast. Who could blame him? Sensory information is the feedstock of comprehension. No one had seen the ice. Everyone had seen floods. It is no small matter to acknowledge that if an idea held dear is inconsistent with reality, that idea must be wrong. But Buckland, in a refreshing display of intellectual openness, eventually reversed his position and became one of Agassiz's strongest supporters. That it would take another twenty five years before the scientific establishment fully accepted the notion that vast sheets of ice once flowed across the land suggests that human understanding is born of multiple parents, that sensory observation alone is not the sole deliverer of perceived truth. Do we believe something when we see it or do we see it when we believe it? We now understand Agassiz's ghost ice sheets to have carved the landscape of the northern latitudes, giving birth to several million lakes in the process.

. . .

I choose to start my search for Lake Agassiz's ghost by looking first for something solid, something my senses and my toes can press against. The lifetime of a lake is marked between its first beach and its last. I am looking for Lake Agassiz's first beach, its highest, the Herman Beach. I want to walk the strand line that tells of the birth of a lake. Directly ahead of me lies Maple Lake, my reference point. The Herman Beach is supposed to bend sharply east near here.

The beach supposedly lies at approximately 1,190 feet elevation. My contour map tells me I'm close. Contour lines also tell me I'm 130 feet higher in elevation here than where this beach line lies a hundred miles south. Whoa. That makes no sense at all. Lakes are flat. Beach lines have to be level. How can a beach be at greater altitude at one location than at another? Surely the lake and its beach did not tilt uphill.

Mr. Agassiz's ghost ice sheet can explain. As the ice thickened and flowed from the ice machine in northern Canada, the weight of the ice pressed down, ultimately causing the land beneath to sink, much as a thick book laid on a foam pillow depresses it. The greater the weight of the book, or ice, the greater the downward compression. When the book is removed, the pillow rebounds to its previous shape. But an unabridged dictionary placed on one end of a foam pillow depresses the foam far more than a Tom Clancy novel placed at the other end. When the two books are removed the more compressed foam beneath the dictionary has more rebounding to do than under the novel.

Likewise with removal of ice. Thicker ice here than a hundred miles south means greater rebound here. The shore of Hudson Bay has rebounded 600 feet and, beneath the ice's epicenter in northern Canada, the land has rebounded an amazing 900 feet since glacial times. Depressed by the weight of its ice cap, most of Greenland's land surface actually lies beneath the level of the sea.

. . .

Glaciers do two things well: they scrape and pluck a landscape bare, then carry away the accumulated rocky debris and leave it somewhere else. Both create basins. Most of the scraping occurred across Canada and parts of the northern United States. Here the unrelenting ice gouged out hundreds of thousands of bedrock basins. Climatic warming eventually stalled the ice, then began melting it back, uncovering those bedrock basins which became lakes.

Glacial erosion of bedrock has produced some of the most beautiful lakes on the continent. New York's Finger Lakes and those of the Adirondacks, most of the lakes of Maine and Minnesota's Voyageur's National Park and the Boundary Waters Canoe Area, and the glorious lake districts of Quebec and northwestern Ontario all owe their existence to the plucking and scraping of bedrock by the relentlessly moving ice.

Like many other lakes, Lake Agassiz was created by deposits of glacial debris carried away from these bedrock basins. About 12,000 years ago the retreating lobe of the glacier stopped receding at what is now the head of the great valley within which I lived as a child. Although the *edge* of the ice became stationary, the *ice itself* kept flowing, but at a rate offset by the rate of melt. So, the ice tongue continued to deliver its mixed load of grit and rock, much like a conveyor belt in ultra-slow motion, its debris accumulating in a pile of glacial till, a moraine, at the melting edge. Meltwater pooled between this ridge of till and the edge of wasting ice to the north and Glacial Lake Agassiz was born. As the melt continued and the ice front resumed its retreat, the lake expanded northward into newly uncovered lowlands.

Glacial Lakes Missoula in the west, Albany in the Hudson-Champlain Lowlands, Lake Hitchcock in the Connecticut River Valley, and even Glacial Lake Cape Cod formed in similar fashion. Agassiz dwarfed them all. As it grew in size and depth, it rose higher and higher against the dam of till. Then, like the sea surge of a hurricane tops dikes, the lake's waters finally topped the moraine dam, creating an outlet and pandemonium. Torrents of liberated waters cut through the till, gobbling up all but the largest boulders in their path in a headlong escape from the lake. Across the landscape the roaring deluge raced. Excavation of the great valley had begun. As the downcutting of the till continued, huge boulders, too large to be moved by the onrushing waters, accumulated on the floor of the outlet. This boulder pavement soon stopped further downcutting. The lake level now stabilized, the Herman Beach began to form.

. . .

I start the car, turn left, and begin driving south. The Herman beach ought to lie on a north-south-tending line of land that slopes west. I scan the countryside. Rolling farmland, scattered potholes, groves of trees—the landscape is much more complex than I expect. My untrained eyes see absolutely nothing that resembles a beach anywhere. It now seems the search will be more difficult than I imagined. This beach has stood abandoned high and dry for thousands of years, time enough for creeks and erosion, farming and gravel removal to recontour the land, obscuring the sign of a beach.

I stop the car to ponder my situation. I locate myself on the map and determine my elevation to be a bit greater than 1,190 feet. I decide to continue driving south and to turn west every few miles on gravel township roads until I intersect an identifiable beach.

I take my first turn west. The gravel road passes beside black farm fields, green shoots not yet up. Soon after my turn the road begins a gradual descent. I slow down. In an instant it is there—suddenly, unmistakably, as plain as the nose on my face. Next to the road, a small bare field, black as dirt can be, scattered stones on its surface, abruptly ends in a wallow of golden brown sand. I stop the car and walk the few steps to the demarcation line.

I scoop up sand in a small plastic cup, roll a pinch between my fingers, then spread it over the palm of my hand. My magnifying glass reveals most of the grains are like bits of opaque glass with slightly rounded edges. They are nearly all the same size. Quartz bits, the makings of sand and glass. Grains of black and pink are sparsely scattered among the particles of quartz.

I turn and take up a handful of the black earth and squeeze it. The mixture of pebbles and sand and particles of clay contrasts sharply with the uniform grain size of the sand. Till, geologists call it, the unsorted leavings of a glacier.

Rocky till, unsorted by water. Sand, uniform in grain size. Side by side, edge to edge. There is no transition. I stand precisely at the upper edge of the beach, where even the storm waves of the ghost lake no longer had the energy to stir and sort and segregate particles. I did not expect such clarity, such definitude.

The wallow is only a few dozen meters long and half that wide. Brush, grasses, and small trees cover the sand on three sides. The beach sand is laid bare here only because the land surface has been disturbed. Some appears to have been removed. Did the farmer need sand for fill, to make cement, for a child's sandbox?

I take another handful of sand and let the grains run slowly through my fingers. From the edge where black meets tan, I look west, down slope, out onto the ghost lake's imagined waters. Though climate change has long since swallowed the ice and this lake, I listen to waves. I can see them, feel them, believe them. The ghost lake is ghost no more.

I stand on the touchstone to beginnings, to the birth event, not just of a beach or even a lake, but to a landscape of half a continent. The water in waves that hurled their energy upon this beach, that sorted and rounded the sand grains in my plastic cup, came directly from the ice that midwifed the birth of a landscape. Place is time and time is place.

Beaches are born of conflict, offspring of the energy of wind and wave hurled in violence or lapped in gentle submission upon the land, the nature of the creation dependent on the depth of the water, the fetch of the wind, the direction from which the waves attack, and the length of gestation.

Waves energize the shore, stirring and tumbling clay and sand and small pebbles. The more energized the waves, the larger the particles that can be moved. Little energy is required to suspend the tiniest bits of clay and transport them to the quieter deeper waters away from shore, where they settle out as mud. Larger particles, too heavy for such transport, are left on the shore to bang against each other, chipping and rounding, to form sand or gravel or cobble beaches. The longer the time of contact between water and land, the more sorting occurs, and the more uniform and rounded the particles become at any one location: fine sand here, tiny pebbles there, stones somewhere else. The greater the force of the waves, the more prominent the beach becomes. With a fetch here of 250 miles from the northwest, a wind of 30 miles an hour could produce eleven foot waves and a beach fifteen feet high.

. . .

Subsequently, increased volumes of outflow from the lake reenergized the outlet river, enabling it to break through the boulder pavement, and the lake level began to drop. After twenty feet of downcutting, enough large boulders had again accumulated to restabilize the outlet and allow another beach, lower than the first, to develop. Again and again the process repeated, creating a series of beaches at progressively lower levels, like the accumulation of bathtub rings as the water is let out of a tub in stages, turning more and more lake bottom into dry land. I must go out onto its bottom to look with fresh eyes, to see.

. . .

I return to the car and drive down the slope onto the old lake bed. What the maps of this country lack in contour lines they make up for in ruler-straight parallel blue lines running west: County Ditch Number Thirty-eight, Number Forty, Number Forty-one . . . Judicial Ditch Number Seventy-five, and so on, all directing water west to the Red River of the North. Rows of tiny plants, sugar beets I expect, extend far into the distance in soil black as black can be. Occasional shallow potholes remain as the most direct reminder of the landscape's watery past.

Lake Agassiz's legacy is both blessing and bane to farm families. The deep rich black clay soils formed out of the lake bottom sediments are among the richest in the world. But lack of topographic relief greatly slows movement of rain water and snow melt off the fields, despite the assistance of hundreds of miles of man-made ditches. Even when the water does reach the Red River, its only way out of this pancake land, the water does not hasten. The river wanders this way then that, searching for the way downhill. In places on the Red a paddler moves seven miles east and west for every three miles gained north. Severe flooding and drowned crops are common.

Vast raven-black fields extend in all directions, uniformity broken only by scattered farm groves, tree islands that become less distinct with distance, then finally float above the surface as mirage, as on a large body of water, before disappearing into open horizon. Flocks of gulls rise as one to circle then drop back to the ebony earth.

At the Dakota border I turn southeast onto Highway 9 to begin driving a zigzag line south to the moraine dam that gave birth to the ghost lake. The second town I come to is Campbell, population 241. A town of grain elevators, storage buildings, and vacant storefronts, it is the namesake of the vicinity where Agassiz's great final beach was first discovered, a beach I hope to find.

I continue out of town. A handful of dirty-white pelicans crowd onto a narrow brown spit of a nearly dry pothole, the birds finding respite on the shore of a pond not yet captured by the thirsty drainage ditches. I pass through Tintah, population seventy-nine, and Norcross, population fifty-nine, and finally the larger Herman. Storefronts vacant, buildings boarded up, these towns struggle to stay alive in a changed world, like shallow remnant ponds in a drying, disappearing lakescape. All will survive, if only in the glacial geology books as the names of prominent beach stages of the great ghost lake.

· · ·

I reach the Army Corps of Engineers' historic site in Browns Valley where Lake Agassiz topped the moraine dam and gushed away as the raging Glacial River Warren. I drive west and climb the bluff that is the South Dakota side of the moraine and look back across several miles of valley to the bluffs on the Minnesota side—the floodgates of landscape history.

The rushing water removed everything down to bedrock behind the Morton school, leaving the scoured kettle holes as its "Kilroy was here" signature. The oversized valley trench, the hills, the outcrops of ancient rock—a landscape created by the outrushing waters of a lake.

At Brown's Valley the river finally reached the ancient undefeatable bedrock, making further downcutting impossible. A long period of stability ensued, enabling the lake's grandest beach to develop, the Campbell Beach, traceable for 1,500 miles.

Though I see several interruptions in the flatness of the land near Campbell, it is not clear to me which, if any, are the Campbell Beach. I turn north to seek it there and to visit Mother in her nursing home.

· · ·

I drive into Buffalo River State Park at dusk to camp for the night, intending to search tomorrow for the last, lowest, and most highly developed beach of Lake Agassiz. I stop at the entrance station as the attendants are about to close up for the day. After receiving my campsite assignment I tell the park ranger of my plans to hunt for the Campbell Beach the next day and ask if he can help me find it. "You're on the Campbell Beach right now!" comes the reply. "Although much of the landscape here has been changed by sand and gravel removal, I can show you where the Buffalo River has downcut through an undisturbed part of the old beach." He marks the place on a map.

A short walk through the woods the next morning brings me to the bend in the river marked on my map. A layered erosion bank stands across the stream. I remove my shoes and pants, hide them in a clump of grass, slip into a pair of running shorts, and step into the water. Sandbars and squishy mud-bottom pools play peek-a-boo among huge boulders, far larger than the river can move. Boulder pavement. The stream bed here has become the type of erosion-proof bed that stabilized Agassiz's southern outlet so many times.

I clamber up the side of the exposed face. Bur oak, basswood, and ash grow out of black soil at the top. A distinct layer of fine sand lies next below. My hand lens shows the grains are much more rounded, more like tiny imperfect marbles, than sand from the Herman Beach. A layer of gravel is next, then comes glacial till, rocks, and stones of all sizes embedded in a matrix of sandy clay, unsorted by moving water. The till disappears under talus at the water's edge. The well-rounded sand grains and pebbles and the prominent lines of this beach attest to the great energy delivered by the waves and the lengthy gestation of this, the most impressive of Lake Agassiz's beach children. The Campbell Beach, at its several levels, was the last beach to exist before the wasting ice exposed a lower outlet in Canada, stealing the River Warren's water, draining away much of the lake, and leaving the miniscule-by-comparison Minnesota River to meander dazedly through the loops and sloughs, oxbows and farmlands of the Warren's giant trough. Outflow from the lake stabilized at fifty-five different elevations during its lifetime, long enough each time to establish a beach.

In the late morning warmth, I pack my tent and duffel in the car and slowly drive the park road back to the entrance. Grasses now cover a terrain that has given up much, but not all, of its beach front to sand and gravel merchants. I stop at a large glacial erratic marking the park entrance and get out of the car. Eight feet across and nearly four feet high, the pink and black metamorphosed block must have had a long ride on a glacier. A small attached metal plaque declares:

CAMPBELL BEACH, THIRD STAGE
Erected by the Lake Agassiz Chapter of the D.A.R. 1933

I leave the lichen-covered traveler and turn onto the highway, my back to pool-table flatland, and head east. U.S. Highway 10 here passes through vast farm fields where the contours are bared. I count eight beach terraces before I reach the rolling hilly moraine topography that marks the edge of the lake country where my parents live. Dad told of a visitor to the area who chartered a ride in a small plane to see what it looked like from the air. The man came back astounded. "From up there it looks like the whole landscape is water. Makes you wonder how people down there keep dry." Lakes large and small, deep and shallow, some with stream flows, others in seepage basins cover the landscape.

By scraping the land bare in some places and leaving piles of rocky debris in others, by abandoning blocks of ice of all sizes in till piles to melt and become lake basins, and leaving till dams backing up lakes, the ice sculpted the face of the land from Glacial Lake Missoula north across northern Iowa to Maine. The vast majority of the lakes we love are simply gifts from the glacier.

Ghost Lake Agassiz comes to life for me thousands of years after it vanished. Ghost ice sheets become real, emerging out of astute sensory observations and reason by Agassiz. How do we take the measure of human perception?

Agassiz moved to America and Harvard in 1845 as conquering hero, Father of the Ice Age, where he further enhanced his stature by establishing an extensive zoological collection and museum. Thoreau was one of his field collectors.

What does it say about the human mind that Louis Agassiz, though able to perceive ice sheets that others could not, was unable to perceive another new idea of his time advocated by Charles Darwin, evolution through natural selection?

Agassiz eventually watched his immense scientific credibility crumble when his perceptive powers could only vigorously oppose Darwin's grand idea long after the scientific establishment went Darwin's way on evolution, as it had once gone Agassiz's way on the ice age. How fitting that the self-described nonscientist Henry Thoreau quickly perceived the truth of both ice age and evolution.

· · ·

I continue the twenty-odd miles east to my mother's nursing home. Automated doors open as I near and enter. The shiny gray waxed floor tiles lead me down the hall where I find Mother in her wheelchair. She is quiet with a vacant expression on her face. Her mind robbed by Alzheimer's dementia, she has not recognized my wife, her grandchildren, nor me as her son for years.

I tell her the family news, the comings and goings of her grandchildren, and the health of her elderly sisters. She listens attentively. I do not know that she understands. I do not know that her ghost mind does not.

I run out of news and must go. I hug mother and wish her goodbye. I have learned that I cannot speak about the lake of her girlhood. Any reference to that lake and her mind snaps into action and she begs to be taken "home." Even Alzheimer's can't erase her memories of lake.

On Seeing

The eyes may trick us into a sense of mastery, but the ears know better. Sight insists on separation; hearing, like touch or taste or smell insist on connection.

SCOTT RUSSELL SANDERS

We say, "Open your eyes and see." It may not be so simple. Here is what I've discovered.

The National Wilderness Act, although a major tool for protecting pristine natural places, defines wilderness, and so defines places worthy of preservation, as "an area of undeveloped *land* affected primarily by forces of nature." The act might have said "land or water." The aquatic systems within the nation's Boundary Waters Canoe Area Wilderness do, of course, receive de facto protection, but explain this: had someone proposed introducing a nonnative flower or shrub species to the wilderness forest because of its gorgeous blooms or exceptional fragrance, wilderness advocates would surely have vigorously opposed the proposition as antithetical to wilderness values, a blight on the native forest. In fact, a program is now under way to uproot and destroy plants not native to the forest that have invaded on their own. Strangely, this ethic did not apply to the waters. Some of the strongest, most effective advocates for preserving the Boundary Waters Wilderness struggled across portages with pails heavy with smallmouth bass fingerlings to introduce into lakes outside the bass' natural range, with negative consequences for native lake trout.

In the mid-1960s, Minnesota began a farsighted effort to "preserve and perpetuate the ecological diversity of Minnesota's natural history" by establishing a statewide Scientific and Natural Areas System. Paradoxically for a state famous for lakes, the program was organized exclusively around terrestrial plant communities. By 1999, 130 prairie grassland, deciduous woods, and coniferous forest sites in the state had been designated. There were no other categories. The several small lakes located in a few of the parcels seem incidental inclusions, not part of a systematic effort to protect lakes. Yet lakes contain biological diversity, are an important part of the state's natural heritage, and are as significant for recreational, scientific, and educational purposes as a prairie or woods. Since lakes can neither be cut down nor plowed under, in ordinary circumstances anyway, perhaps they were considered immune from risk, not in need of protection.

Even major national environmental organizations dedicated to preserving remnants of diverse ecological communities have focused almost exclusively on terrestrial settings, despite acknowledging that freshwater systems are the most endangered of all. Lake protection has been notably absent from their efforts.

My puzzlement turned into understanding one day while working with students in the most unexpected of places, a mature maple-basswood forest not far from Diamond Lake. The place is called Taylor's Woods. I bring my students there to help them learn important ecological ideas. We park cars in a grassy space next to the county road and hike less than a quarter mile on a horse trail to the wood's south edge.

For a width of several yards the edge of the woods is a mass of raspberry brambles, sumac, sedges, and a crowd of pencil-thin saplings of ash, maple, and cherry that tugs at our clothing as we walk through. Past the edge, magnificent maple and basswood pillars hold up a backlit green ceiling high above.

After describing a brief history of the woods, I explained to my students the assignment for the day. Once all were busily engaged, I scouted for things to show them once they had finished gathering data. An expanse of wood nettle gave way to a knot of trillium here and a mat of bedstraw there.

A brilliant scarlet cup, the fruiting body of the fungus *Peziza,* peeked through a gap in leaves dropped the previous fall. But none of my students have seen the bright red midge larvae in lake bottom mud. Delicate thread-like fungal strands wove through rotted brown material—no longer wood but not yet soil. Large cavities left by a pileated woodpecker hungry for beetle grubs scarred standing dead trees. Centipedes, mites, beetles, spiders, millipedes, and other forms of crawlers and wigglers scurried for cover from beneath an overturned rotting log. But vanishingly few have seen the rich assemblage of micro-crustacea beneath a lake's surface or the tullibee who eat them.

Students joined me as they finished their work and we walked on together, drawn into the lives of the woods' inhabitants. A sea of sugar maple seedlings spread ahead of us, tiny sprouts so dense it was hard to avoid stepping on them. They were uniformly one inch tall. Seed coat remnants still clung to thin, tentative stems emerging through the moist layer of leaves. Why was this the only age class of maples on the forest floor? There were no ten inchers, none of brush size. Where was last year's group and the one from the year before that? Students quickly decided a dense island of brush ahead of us provided the answer. But none have seen the acres of underwater gardens of algae as large as blueberry bushes, or the green arbors of freshwater sponges or their response to the coming of winter.

Taylor's Woods has an open interior, in contrast to the thick brush common in other woodlands. We walked closer to the brush island, stopped and looked up. The ceiling had a hole. A fallen maple matriarch had left the opening, allowing light to brighten the forest floor. The stored food in maple seeds can support seedlings for only weeks, then the youngsters must feed themselves. Unless a light window opens above, each is doomed. The silent peace of the crowded island of plants belied the intense competition for light raging among them. My students readily concluded that the island was large enough for only one, maybe two mature trees to ultimately reach the canopy. Early death awaits all the rest. Light is clearly the great arbiter of plant life, molding the character of the plants and the physical structure of the woods.

. . .

On one such class visit I became conscious of the many observations and explanations I was sharing with my students. I also noted we were probing more of substance than simply names of flowers, trees, and animals. Why, I wondered, should I, as a trained aquatic biologist, see so much to share with students in a *woods,* for heaven's sake? We pressed on.

Though maple seedlings were common, we saw none of basswood, despite mature basswood trees all around. Maple seedlings tolerate shade better than seedlings of any other tree species likely to compete with them for light. But basswood has a formidable trick of its own. Wherever a basswood has fallen to earth, suckers sprout from the tree's base. Fed by food reserves in the massive roots of the prostrate veteran, these sprouts grow explosively toward the sun, rapidly outdistancing scrawny seedlings of any type that might be nearby. This feature of the basswood lifestyle accounts for its common growth form as a clump.

Tan shelf fungi clung to the softened wood of a prostrate maple taken down by strong winds years before. Moss grew over the trunks of others half buried in a bed of black and brown. Their rotten trunks sank spongelike underfoot. Once rock-hard wood crumbled in my hand. A crowd of maple seedlings and a white-petaled trillium grew out of wood turned into soil. From life to death to life again the grand material cycles turned, an open secret in the woods.

A student noted we are intimate participants in the life of the woods. Our exhaled breath provides the carbon dioxide with which a maple tree makes sugar to create new leaves, twigs, and next year's crop of seeds. The oxygen released from the maple's leaves fuels our cells enabling our every movement, thought, and sensation. So much happens in plain sight. By opening their eyes, my students perceive.

. . .

When I bring my students to experience a lake, the contrast is stunning. Diamond Lake nestles in a depression among gently rolling hills less than a mile beyond Taylor's Woods. The smell that says "lake" announces its presence before we reach the water's edge.

The touch of breeze through the hair, the *konk-la-ree* call of the red-winged blackbird and the swash of the waves provide a sensory-rich greeting on our arrival at the shore. A green heron often flies off at our approach. Cattails and rushes and floating green patches of coontail, elodea, and duckweed fringe the water's edge.

Tiny bits of life occasionally zip between plant stems sheathed in a filmy layer of yellow-brown and disappear. Random gulls and a covey of coots share the open lake. But shore and surface are merely the lake's edges, like the dense band of saplings that marked the edge of Taylor's Woods. We have not yet penetrated beyond the lake's perimeter. (Barbara Hurd reports that Monet vowed that trying to depict what's under water almost drove him crazy.)

I can enhance my students' experience by handing out nets and scoops to sample shallow waters for tadpoles, insect nymphs, and assorted crustaceans, fascinating creatures all. We can go in boats to midlake, where the Eckman dredge brings bottom muck and rich collections of blood worms and larval midges to the surface and to our consciousness.

I can expand my students' perceptions of lakes by having them collect water samples, adding the powder contents of small plastic pillows with which, they are told, they can determine the milligrams per liter of oxygen dissolved in the lake water. My students know about oxygen, of course, but much that they know does not apply to life beneath the waves. Stand atop earth's tallest mountain and you can experience oxygen deprivation, but nothing like that experienced by life underwater. No, oxygen is neither plentiful nor in consistent supply, and sometimes there is none.

The closest my students can come to understanding oxygen in lakes is to view a row of oxygen sample bottles back in the lab arranged in the order of depth from which each sample was taken. Chemically fixed oxygen molecules produce amber-colored solutions. Bottles containing samples taken from many lake bottoms in midsummer are devoid of color, the sign of oxygen's absence. Samples taken from the same place in the fall produce a rich, amber color, the color of urine produced by a dehydrated body. My students can translate those varying shades of amber into intellectual understandings. But oxygen as amber liquid? Experience tells them otherwise. Electronic probes now produce oxygen numbers far more quickly than the chemical method, but even more abstractly as a digital readout, further reducing students' perceptual experience.

We can supplement our limited perception of lakes by intruding, momentarily, with snorkel mask or SCUBA, or sitting in a darkhouse on winter's ice staring transfixed by the brown-green world below. The blurred streak of a northern pike's ferocious attack on a fish dangling at the end of a line enriches anyone's knowledge of lake. But these are fleeting experiences, brief looks, as ephemeral as morning mists that fade with the rising sun.

Human understandings rely heavily on the endless flood of messages from our eyes. But with a lake, the laws of optics work against us. Its reflective surface and the water's rapid absorption of light with increasing depth conspire to keep much of a lake hidden from view. At a lake, open your eyes and you don't see half of it. Nearly all the visual input from a lake comes as surface reflections, mere superficialities. Thoreau reveled in reflections. He felt because they were constantly changing they could present previously known truths from new points of view. But those truths don't penetrate into the world of the lake below, a truth with which he seems unconcerned. What kind of landscape even exists in the absence of vision? How can we fully perceive what we cannot see?

We cannot live with the sunfish, skin by scale. We cannot accompany lake herring in the tenuous days of summer, progressively squeezed between oppressive, unsurvivable heat in the waters above and lethal oxygenless waters expanding up from below, where life's choices become asphyxiation or heat stroke.

It is as if a lake were a metaphorical theater. I have often taken my seat at this theater, on a log at the shore or in a canoe on a lake surface. But for me, as a creature of the land, the curtain never truly goes up. It remains closed but for a snippet of monologue here, a moment of repartee among a few members of the cast there, that whispers out through tiny scattered rents in the curtain. The rest of the time I can but wonder about the truths in the watery depths, wonder about life behind the curtain's veil.

Hastening Slowly

What is it that we are not seeing?
What is it that we are not hearing?

TERRY TEMPEST WILLIAMS

Of the available means of travel from one watery place to another, the canoe, with mile made stroke by stroke, yard by yard, must surely be the slowest. While that may be true for miles covered, what if the destination is to understand, to establish dialogue with a lake? Measures of speed depend on one's destination.

I discovered this truth my first time in a canoe during summer break from college at the invitation of a friend. Pat asked me to join him for a weekend canoe trip on a lake now part of Voyageurs National Park. I had traveled the border lakes by motorboat for two summers but had never set foot in a canoe. I readily agreed to go.

Pat's family cabin, near the mouth of the Ash River, served as our jump-off point. Packs filled with tent, sleeping bags, and provisions sat assembled on the dock. Pat retrieved the canoe from beside the cabin and set it in the water. The green canvas-covered wood craft bobbed light as a leaf alongside the dock. Its sleek symmetry, simple design, and gently upturned ends spoke of a heritage little changed over the millennia, a craft born into an understanding of lake. Tiny cracks in the paint and well-worn seats revealed this to be a seasoned veteran of lake outings.

As I began to step aboard, Pat warned me, "Canoes are a little tippy, step onto the midline." I stepped onto the wood ribs and the canoe wiggled as though it were alive. I took my paddle and settled into the bow seat. After explaining the proper way to hold the paddle, Pat demonstrated the forward stroke and draw stroke, and we were off.

The mists of time blur most of the details of that first experience. I don't recall if we caught fish or even if we tried. I remember little of the weather, the precise route we took, where we camped, number of miles covered, or even our destination. I do remember unpracticed arms so sore they groaned for mercy before the end of the second mile.

Most of all I remember sensing how the canoe put me into a new relationship with the lake. I sat much lower and, with the narrowing at the bow, so much closer to the water than when in a boat. My lower hand on the paddle could directly touch the waves. I could wet my fingers in their coolness. I could dip a cup of water and drink. I learned that with practice I could even point a dripping paddle skyward and let its water run down the blade and feel at least some of it dribble into my mouth.

Fully loaded, the canoe drew less than three inches of water, allowing us to slip easily through beds of reeds and lily leaves. We explored shore edges and shallow bays where no motorboat could possibly pass. I watched amazed at the countless insects large and small that walked or hopped on the surface of the floating plants at my fingertips. The distinctive smell that distinguishes lake from land seemed so crisp, so penetrating, when sitting at its source, undiluted by a motor's fumes. Though not *in* the water, I was surely *of* the water.

· · ·

Pat's plan called for us to leave the lake on which we began some miles after launching and carry our belongings, canoe and all, to another lake and to a wonderful realization: With a bit of help from us, the canoe could take us wherever we wished. The craft freed us from inconvenient geography. How liberating!

I also discovered silence, silence that enabled me to hear the wind in the rushes, the lap of water against the skin of canvas, the calling frogs. For the first time I felt the pulse of a lake.

By slowing us to the pace of a walk, Pat's canoe freed my senses to inhale the lake experience, to flood my mind with *lake,* to feel an intimacy I had never known.

LAKESCAPES

Edges

A man sees only what concerns him. . . . How much more, then,
it requires different intentions of the eye and mind to attend to
different departments of knowledge. How differently the poet and
the naturalist look at objects.

HENRY DAVID THOREAU

Some might describe the distinctive odor of a lake as smelly, even obnoxious. Except when the stench of rot pervades the shore, for me lake smell is alluring, an invitation that draws me like a pheromone.

Geri and I are camped in a state park campground a hundred yards back in the woods from a large lake's edge. We can smell the lake from our tent. I am enveloped by that smell as I sit on a boulder inches from the water.

I have come here to immerse my senses in this shore, to follow eyes and ears and nose and fingers wherever they lead and open new "departments of knowledge," as Thoreau put it. Edges intrigue me. Is edge the place one thing begins and another ceases to be? Or is it where two things blend to become one?

From my boulder, the line where water meets land traces a long gently curved crescent into the distance, cleanly marking the divide between one world and another, as clear a separation as a line drawn in the sand, where my toe gets wet on one side and remains dry on the other. At least that's the way it seems.

Here the line is edged with stones. Nearly all the rocks are large, many over two feet across. Sluggish waves flood into the chambers among these boulders to melt peacefully away in nooks and crannies. Waves here are not always so passive. Geri and I once brought our newly purchased sea kayak to this lake to test its seaworthiness. A strong northwest wind drove great waves out of the open horizon that day, hurling them toward land. Huge swells toyed with our twenty-foot craft, then threw themselves crashing onto the shore. Windrows of driftwood and bulrushes decay in the trees behind me, testimony to the great energy with which waves can attack the land.

Northwest winds and the waves they produce strike the shore obliquely here. The considerable energy the breaking waves discharge to the shore does not die in the foam of the rocks, but is transformed into a mover of particles—sand grains, tiny pebbles, and, on this beach, even cobbles, leaving only the largest of rocks behind. Today's waves are too puny to move much of anything.

A dozen pinhead minnows hover above a flat, submerged rock streaked with bands of red and pink and spots of gray. Waves toss the tiny fish back and forth across the top of the rock in rhythm with the ebb and flood of the water. Rougher waves from a passing boat swirl the tiny fish and they abandon the rock for deeper water off its edges. Long strands of algae sway like a mop of green hair fringing the rock's outer edge, hiding tiny creatures the minnows would love to eat.

I turn over an underwater rock the size of my hand. A mucous egg cluster clings tenaciously to its underside. A small rust colored crayfish emerges cautiously from a crevice, glides slowly forward, and disappears into a crack. Another overturned stone reveals an elongate, flattened creature less than half an inch long plastered against the rock's bottom. Its six stout legs extend spread-eagled, as though clinging to the rock for dear life. Two large eyes dominate the top of its head, and three filaments stick backward like tail feathers. Pairs of leaflike gills along its abdomen reveal its identity—a larval mayfly. Like the great majority of insects, he is safe for me to handle, though it might take tweezers to pry him loose from his grip on the rock.

After many molts he, maybe she, will crawl from the water, shed the last exoskeleton of youth, and transform herself into a subadult, the fly fisherman's dun. In a day's time she will extract herself, gauzy wings and all,

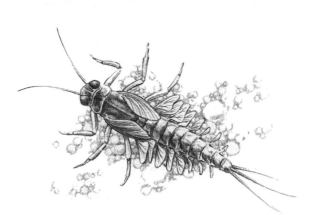

Mayfly larvae use feathery lateral gills to absorb oxygen.

from her thinner-than-paper external skeleton and emerge as a glistening adult, a spinner.

This transformation of a six-legged wormlike creature, adapted to live under water, into a terrestrial creature that flies, surely ranks among the marvels of the lake. My aquatic entomology professor once likened the process to a submarine turning itself into an airplane yet remaining functional throughout the wondrous passage. Imagine working as a pearl diver for most of your life, then, never having seen a plane, being told you are to take over the job of piloting one—starting tomorrow.

The mayfly line has graced the earth for more than 300 million years. Long before there were butterflies or beetles or societies of ants and bees or creatures with hair, there were mayflies. They have existed on earth longer than any other aquatic insect group, preceding even dragonflies by fifteen million years.

The mayfly lineage has eluded seven major extinction events that shook large numbers of species off the tree of life. How can one not admire the mayfly clan? I replace the rock as I found it, returning the young mayfly to scraping the rock of algae, bacteria, and other morsels of food.

A tiny black leech extends its head end across the underside of another rock, winding among a pod of transparent egg cases shaped like miniature

Volkswagens. Stretched to the diameter of thick pencil lead, the leech finds the rock's edge and slips back into the lake. Two quarter-inch-long creatures that resemble miniature crayfish lay curled around each other in a depression in the rock. They wiggle in the drop of water that cradles them. Small fish would gobble them up in an instant.

Elsewhere sand grains are bound together into the shape of a tiny, beautifully coiled snail shell, a portable home constructed by insects known as caddisflies. Larval caddisflies use a powerfully adhesive silk to bind the grains together. Its owner, whom I do not yet see, lives much like a hermit crab, carrying his shelter with him, sallying forth to feed but able to quickly retreat "home."

A swarm of tiny midges, another flying insect that spends its early life underwater, suddenly forms and as quickly disappears. I once mistook a massive emergence of these mosquitolike, nonbiting fliers as smoke from a distant fire. Many spiderwebs span the narrow crevices between cobbles and boulders of this shore, and nearly all hold midges.

I saunter toward a point at the crescent's southern end. Stones are smaller here, no larger than softballs, nestled in tiny patches of sandy gravel. The curving shore has changed orientation to the prevailing wind. This part of the beach is spared the more energetic storm waves allowing smaller stones to remain unmoved.

Past the point, a quarter-mile-long beach of sand forms the next crescent. Halfway down this beach buoys mark a swimming area where a tiny dab of a girl walks into water up to her chest then splashes wildly in pure joy. Three young boys push each other beneath the water in raucous play. All told, a dozen people entertain themselves at the water's edge.

Tiny wavelets caress the beach obliquely, nudging the tiniest sand grains down the shore a millimeter at a time. The waves vary ever so slightly in energy and, at each fourth or fifth wave, larger grains are set in motion. Each differently sized particle requires a different threshold of wave energy to be moved. The moving water is constantly sorting, sorting, sorting, accumulating sand in one place, leaving pebbles in another and cobbles somewhere else. Imperceptibly, the sand of this beach moves down the shore, stopping only when the wavelets are at rest. Today's waves dictate that beach sand creep south. Tomorrow? Different waves, different angles.

Maybe the sand creeps back north. A tiny point of sand protrudes several feet into the lake, interrupting the gradual curve of the beach. Three rocks the size of grapefruits rest at the tip of this tiny point, left by the children. By interrupting the wave pattern with their rocks, the children have stalled the migrating sand, trapping it to form this tiny "point," their personal signature on the shape of the beach.

My family once spent a day creating a small sandy beach by removing stones and pebbles from a cobbly shore on our lot on a small Wisconsin lake. At day's end, the sand that remained became our beach. But waves and wind and orientation of the shore conspired against us, and in a year had washed away the sand uncovering a new layer of stones.

. . .

Hundreds of insects fly erratically among young ash trees at the upper edge of the beach. Many others stand quietly, antennae twitching, or walk on the stems and leaves of the trees. Apparently, they emerged en masse from the lake. Some are a uniform dark brown, with bodies the length of a child's fingernail. Others, twice as large, are mottled tan. Both kinds resemble moths with one obvious difference: they fold their wings over their backs, forming a pup tent–shaped roof. These are adult caddisflies. They have massed among the ash trees, as though seeking escape from the gentle wind. Most of the tan ones walk in a palsied frenzy to the end of twigs, then back again, bumping into each other, as though dithering about getting to a ball on time. Smaller dark ones seem either to be aloft or standing motionless. One has remained on my sleeve for ten minutes.

Spiderwebs entwine many leaves and branches, creating an aerial death trap for insects that fly along the upper edge of the beach. Although the webs have caught many insects, mayflies make up a disproportionately large portion of the spiders' catch. Many mayflies dangle upside down from strands of spider silk, like slaughtered cattle hanging from meat hooks. One long strand holds eight of them in a line.

Their large compound eyes, spindly legs, transparent wings folded vertically over their backs, and two ridiculously long filaments extending from behind appear in death as they are in life. Many remain airborne, still free of the spiders' traps, more by luck than design. These mayflies fly weakly,

Adult mayflies often emerge from water en masse forming dense clouds around lights.

as they float to a landing in the foliage at the head of the sand. Their flight paths show no adjustments in direction, as though they are exhausted, or lack fine motor control, and are content to land on whatever appears in front of them. No wonder they fill the spiders' webs. Death by spider or any other means shortens adult mayfly lives hardly at all. Most will mate within a few hours after emerging from the water, then die.

The air has calmed and a swarm of mayflies has massed offshore, the habit of males. The insects fly upward to the top of the swarm. Then, with wings spread sideways, they slowly glide down through the swarm then zip back to the top, repeating this elliptical movement over and over again. Females, I'm told, enter such swarms then quickly exit with a mate. They copulate in flight, an act requiring much more maneuverability than exhibited by the clumsy creatures coming in to land beside me. Most females deposit eggs on the water, although some species attach eggs to submerged objects.

Mayfly emergence and mating swarms are familiar in lake country, and I've witnessed some spectacular affairs. Once, traveling through a lake in the Churchill River country of Saskatchewan, Geri and I paddled for over a mile through a thick soup of shed nymphal mayfly cases. The cases had piled up many inches deep at the marshy shore. Air can be thick, highways slippery, and towns overrun with mayflies, landing on cars, trees, light posts, people, and every other available surface. They are particularly attracted to lights, under which their soon dead bodies stack up like fallen leaves. Several have just landed on my shirt. How delightful. I now see them in every detail.

I have seen people flip out when these harmless creatures get close. How sad to carry such unleavened dread. Fear of adult mayflies is particularly misplaced. Their mouthparts are functionless. Their sole task is to reproduce. Despite the reputation that bees and biting flies give to the six-leggeds of this world, the vast majority, like the two mayflies keeping me company, are as harmless as jelly beans or blueberries. If we can enjoy butterflies, why not mayflies, caddisflies, dragonflies, and other life with which we share the shore?

Maybe some of our perceptions are predispositions, instinctive, fixed. Spiders and snakes create angst across all cultures, likely a legacy of relationships in our evolutionary past. Are the rest culturally acquired? Though my mother didn't seek the companionship of mayflies the many times we encountered them as I grew up, neither did she view them as frightful or dangerous. She would gently clasp the wings of those that landed on unwelcome places on my sister and me, and gently launch them back into the air from which they had come. At my sister's initial revulsion at a long-legged phalangid, mother was quick to respond. "Oh Mitzi, don't be afraid of him. He's just a daddy longlegs. He won't bite. Here, we can take him by the leg and put him back on his log." Though I'm squeamish about large spiders and cockroaches, to this day daddy longlegs strike me as wholly benign. Our mind's preconceptions of insects appear to lock in attitudes.

. . .

A whiff of air fills my nostrils with a strong fragrance of lake and I decide to experiment. A modest wind blows directly onto shore from open water. I step to the lake's edge and take several deep breaths of air blowing directly into my face. As marine algae emit chemicals into the air, creating the distinctive smell of the sea, life in the lake may too. Steeped in this pure lake fragrance, I walk over stones and sand and windrows of decomposing rushes, straight back from the water to where the upper edge of the shore meets the woods, turn back into the wind and breathe deeply again. Smells wafting in from the open lake now blend with those arising from the shore, producing a more intense organic aroma, the smell now a union of lake and land. Several steps back into the woods the fragrance changes again, dramatically, as new odors of forest molds and woodland flowers blend with smells from the lake.

A swallow swoops in a circle toward the lake, then banks sharply shoreward at the edge of the water, picking off hapless mayflies that have not yet reached the trees.

A huge dragonfly now rests on my right shoulder. What a beautiful creature he is. He must be three inches long from tip of head to end of abdomen, and his wing span must be four. Brilliant turquoise ellipses adorn each abdominal segment. Transparent membrane cells, like colorless pieces of glass in a leaded window, make up his wings. He has immense eyes of shimmering iridescence. He now cocks an eye and wipes it with a foreleg. More elongate strokes of turquoise brighten the body segments behind his head. My shirt wrinkles as I walk, but he seems not the least perturbed, adjusting himself to accommodate my movement.

My neighbor at home, an experienced outdoorsman, tells me he swats at dragonflies whenever they come near, fearful they might bite. How do we arrive at such beliefs? He was surprised to learn they neither bite nor sting. Dragonflies, as insects I suppose, start out burdened with some insect version of Original Sin, forced to face a world filled with, as Eric Hoyt puts it, all the "steppers, swatters, and screamers who live in fear or dread of six legs." Hoyt suggests an explanation for their estrangement. "[Insects] wear their skeletons on the outside, bite sideways, smell with antennae, taste with their feet, and breathe through holes in the sides of their bodies. Their eyes are placid unmoving orbs; when we humans look into them, we experience neither recognition nor empathy." Maybe.

· · ·

Long ago Geri and I visited her Texas relatives. As bedtime approached that first evening, Geri's aunt commented casually that since the pest exterminators had recently made a scheduled visit, we were unlikely to encounter cockroaches during our stay, but that, well, sometimes things happen. "You take the kids' room here," she said. "See you in the morning." Moments later I pulled the bedspread back and a large brown cockroach scampered across my pillow. I recoiled in instinctive fright. How horrible! Though shaken by the intruder, I was even more surprised by the nonchalant reaction of our hosts to this news. How could I, a biologist, find the roach so repugnant when to these people it was a ho-hum event, like seeing an ant cross a sidewalk?

My revulsion at roaches undoubtedly arose at the side of Grandma Matilda. Her word for them was ISH! No cardboard box containing groceries or most anything else could enter her house without first being emptied and quarantined outdoors for hours to make certain any roaches they might contain had skedaddled. Though I never saw my first roach until decades later, Grandma's vehemence taught me that full-scale catastrophe had struck the house if a roach ever got inside. After considerable thought, I cannot offer a single *rational* explanation for my repugnance toward roaches, while I welcome most any aquatic insect that might land on my arm.

· · ·

I first learned of dragonflies by the name "mosquito hawks," because of their ravenous consumption of those troublesome pests. To me, the logic was compelling. Anything that reduced the number of my mosquito bites was clearly my friend. Thus predisposed, I have ever since considered it my wonderful good fortune to have them near, even better to have them use me as a perch from which they can sally forth, capturing mosquitoes and biting flies before they pester me.

Dragonfly flight is without equal. As dragonfly expert Philip Corbet puts it, "On a scoring system that reflected versatility, dragonflies would most certainly emerge as the best fliers this planet has yet produced." Among the first insects to fly, they have had millions of years to hone their skills. They can fly in any direction including backward.

But it's dragonfly vision, unexcelled among insects, that makes their flying skills so deadly.

I look into my companion's eyes. The two immense metallic green and yellow hemispheres dominate the head and allow a near 360-degree lookout on the world. Like the eyes of other insects and crustacea, these eyes are composed of hundreds of individual optic units. These eyes see in color and in ultraviolet and can determine the plane of light polarization. But most of all they are profoundly gifted at detecting motion. The optic units can fire, recover, and fire again over 300 times a second. Were my companion to watch a movie, he would see it as a series of separate still images, an ability with deadly consequences for the mosquitoes and flies he hunts.

Fifteen minutes pass and my passenger still perches on my shoulder. The cloudy sky may explain his disinterest in taking flight. Dragonflies love sunshine. Without it, their body temperatures may fall below that required by flight muscles. Or maybe he is enjoying my company as much as I'm enjoying his. The gray sky may be my companion's good fortune. A swallow flits along the shore just ahead, turning sharply to nab insects on the wing.

I once participated in such an aerial food web on a warm cloudy day by the lake close to my house. A thick troop of mosquitoes fed on me as a dragonfly squadron darted several feet above my head, feasting on the mosquitoes. Higher in the sky, swallows and a pair of nighthawks picked off dragonflies. A patch of sun appears and my rider flies off.

. . .

A mayfly near death makes its final struggles to remain upright on my journal case. Its transparent wings tilt far out of vertical. Legs struggle and the wings again become erect, but only for a moment. The creature totters. Powerless muscles can no longer assist, and the animal falls on its side. A final twitch of a leg and the body is still. A year of life in the lake, a magical transformation to flight, a few hours to mate and lay eggs, and life is through. I gently tip my case and the mayfly slides into a tuft of foliage. I gather journal, pencil, and sitting pad and return to camp to join Geri for lunch.

. . .

Our campground lies about a mile from town. We have run out of milk and bread and decide to walk to the grocery store to resupply. We hike past the park entrance station and follow the main park road back out to the highway and town. Part of the road lies on a causeway that cuts across the mouth of a shallow, vegetation-choked bay. Though part of the big lake, the swampy shores here bear little resemblance to the shore of the open lake. Thoreau wrote, "The attractive point is the line where the water meets the land." The dilemma here is where *does* the water meet the land?

. . .

Viewed from a fishing dock, "shore" begins at the edge of open water in a riotous tangle of pale spaghettilike stems of water crowfoot. White flowers poke everywhere above the water surface. A few yards closer to land these plants are joined by white and yellow lilies. One could neither motor nor paddle through such rank greenery. Thirty yards closer to land cattails begin. Thicker than corn in an Iowa field, the cattails simply swallow the lake in a green sea of leaves. Where does water meet the land? Barbara Hurd describes the dilemma this way. "Trying to define the edge of a swamp is like trying to put a neatly folded shadow into a dresser drawer. Our efforts to outline these places are desires for tidiness, a wish that nothing undefined lurked around our edges. But the truth is, even human boundaries shift."

In other places I have paddled as close to "land" as the thick vegetation would allow, where turtles sun and blackbirds build nests in the crotches of cattail clumps. I have startled green herons from their nests and scared up countless great blue herons in such places that were neither lake nor land. I've left my canoe floating at the outer edge of cattails and tried to slog through them to reach solid land. The outcome quickly became far from certain as each step got me closer to bottom than to shore.

I have ventured into these places in winter, when ice cover makes walking possible. I've followed the tummy trails of otters in snow from one open water hole to another and walked beside tracks of fox and birds and mice and voles. I've discovered unexpected patches of open water hidden within the brown cattail jungle and, more than once, ice too thin to support the weight of a human.

Thick cattails are nearly impenetrable, even in winter. The long strap-like leaves hold you back like a gauntlet of buckled seat belts. Others trip you, sending you stumbling headlong into snow-covered arms of plant clumps. Cattails give way to sedges, then to willows and dogwood, all plants that thrive with wet feet. Farther inland, finally, comes the willowy growth of aspen suckers. Where, indeed, has lake become land?

. . .

I return to the big lake's shore for a final look before we leave for home. Fewer insects fly, though the large darker mayflies still emerge from the water. I walk toward a picnic table to collect my thoughts and make final entries in my field journal. Abruptly ahead lies the ravaged skeleton of a great blue heron. I have walked here several times. Why have I not noticed it before? Disheveled feathers dangle from one wing. The bones have been stripped of what skinny strings of meat they once held. The long spindly femur is light as ether. Besides some ducks, the swallows, a few gulls far out over the lake, and a trio of little brown birds scouring tree trunks for insects, I've noticed few birds on this shore.

I review my notes. Insects occupy page after page after page. An objective reader would conclude I saw nothing else. It appears I have stumbled into the trap Thoreau warned against. Has my fascination with aquatic insects narrowed my perceptual field? Have I given undue attention to the six-leggeds? Apparently so. What about birds and plants? What else have I been too occupied with insects to notice?

My underlying premise for this visit appears flawed. Perhaps the senses cannot operate as open, objective, independent gatherers of information truly unshackled from influences of mind. Can a stomper-swatter come to appreciate the beauty of a dragonfly eye or the fragile intricacy of the mayfly wing so marvelously and wonderfully made? Can one come to appreciate that a fluttering abundance of mayflies and dragonflies and caddisflies at the shore tells of an ecologically healthy lake, one not yet overrun by pollution, if fear interferes?

. . .

Thoreau looks to be correct. My eye and my mind must have different intentions to see other than what interests or directly concerns me. Understanding human perception seems no easier than defining the edge of a swamp. It's hard to grasp a shadow.

Yet, my unintended narrow focus on insects has changed my understanding of the shore. The boundary between lake and land now seems less a line of separation and more a place of union, a grand coming together. The beach itself is a joint creation of lake and land. Aquatic insects feed fish in the water and birds on the land. That these insects, intimate members of food webs in both places, reproduce on land yet rely on water to produce their young shreds the notion of separateness. A lake edge extends past where water first encounters something dry. One cannot place a single "neatly folded shadow" into two separate drawers.

Lady Daphnia's World

*At some point you are seeing so intensely that you become
what you see, you merge into the drop of water until the "you"
disappears. The how and whys and wherefores disappear too. Yet
when you emerge you are somehow replenished.*

PADNA HEJMADI

I sidle up to the edge of cattails, settle myself on the bottom of the canoe, and lean over the side to get my face as close to the water as I can. I am pushing the season by coming in early May. Populations of the creatures I seek normally peak in June, when the water has warmed. Aquarium aficionados and students know my quarry as waterfleas or daphnia. Lake biologists know them as cladocera (cla-da'-ser-a), tiny kin of lobsters and crayfish. I know them as fascinating creatures in lakes that first endeared themselves to me in college days. I also know them as members of the plankton, that vast assemblage of algae, tiny animals, and bacteria freely suspended in lakes.

In our world of mega and supersized, things as small as plankton don't project much significance. Can objects so small really matter? Don't doubt it for a moment. A lake without plankton would be like an orchestra without violins. A lake without plankton would be a lake without fish. Yet barriers to knowing the plankton are formidable. Nothing in our experience in a terrestrial world compares to plankton. Most plankters are very small, visible as specks of green or bits of movement. Worse, they are submerged, making them difficult to see regardless of size.

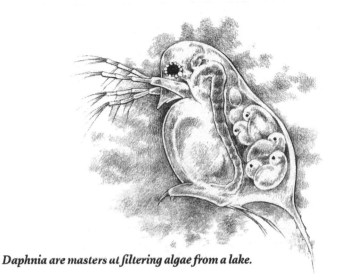

Daphnia are masters at filtering algae from a lake.

Waterfleas are among the largest of planktonic animals, and the largest waterfleas likely to be seen are a group known as daphnids. They can provide a window into the plankton world. This is my first search for daphnids in Diamond Lake.

The lake is calm this morning and the water notably more clear than in my mind-bending visit in August eight months ago. My Secchi disc reveals water clarity at midlake at two and a half feet, over six times as clear as it was then.

The greenish cast to the lake today reveals plankton, plankton everywhere, the vast majority are blue-green algae. Cladocerans would be difficult to find among the dense blue-greens. I will look for them instead in my favorite nook of the lake, a southwest bay that receives pleasantly clear water from a small inflowing creek. Lakes snuggle up to the land most intimately in small bays protected from prevailing winds. This is such a bay. Sheltered by the bay's shape and strategically positioned cattail stands, inhabitants of its waters escape the pounding chaos delivered by waves and wind from nearly all directions. Daphnians should thrive in such quiet waters.

Animated bits of life flit between plant stems. I scoop water quickly with a bottle and peer in. Cladocera can be as large as small ants, and I recognize them in my jar by the jerky flealike way they propel themselves through the water. Counting them is difficult. All are in motion. I judge there to be ten. Most are young. A single large daphnian with pastel salmon tint ascends in slow hops.

No matter how often I dip these bits of life from lakes and ponds, the fascination is lessened not one whit from the first. I am no different than my students when they first see these creatures with a bit of magnification. I can only guess at what incites our passions.

Feathery legs flail inside an open chamber. Antennae stroke vigorously downward, propelling the creatures through the water in their characteristic hopping style. Explosive extensions of springlike legs kick others along the jar's bottom. They appear much more complex inside than seems possible in an animal the size of a kernel of grain. Cladocera are disc-shaped with bundles of organs stuffed inside rigid taco shell-like coverings. The transparent shell displays a beating heart, a pulsing intestine, and babies repositioning themselves in a crowded brood chamber. A large single black eye in continual motion dominates the head. Who would expect these organs of ours inside such a grain of life? Maybe it's the astonishing complexity in such a tiny package that captivates my students and me.

"Oh, Dr. Nelson, come look at this! Is this the heart? It's beating!" Zoologists may view the heart as merely an animated pump, but the irresistible pulsing mesmerizes my students. The rhythmic beat touches a deep chord in some psychological language, that our words—heartwarming, heartfelt, heart of the matter—reinforce.

What must life be like for so small a creature submerged in the waters of a lake? Though we can never know the experience directly, we can do so vicariously. Imagine yourself transformed into a daphnian the size of a tiny seed like the one cavorting in my jar. Now submerge yourself into the lake.

Notice you are slowly sinking. Yes, you are denser than water and even in a lifetime of trying are doomed to fail at the dead-man's float. Take a couple of strokes with those leglike appendages protruding out of your head where your ears ought to be to stop your descent. Feel strange? Your stroke jerked you only a short distance forward before your movement stopped cold. I should have warned you. At your tiny size you have entered a world

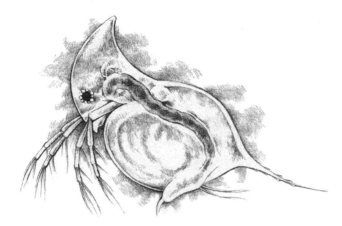

Daphnid helmets make it more difficult for prey to swallow these creatures.

where friction matters far more than gravity. Swimming through water as a daphnian is like swimming through molasses as a person.

Uh-oh. A tiny minnow approaches. Quickly, stroke your way to cover in those plants. Maybe he won't see you. All tiny fish, and some large ones too, love the taste of cladoceran. I learned that one semester at the college. In a mix-up in ordering, we found ourselves with six extra culture bottles of daphnia not needed for class. I emptied them into a large, shallow aquarium well stocked with green algae. The daphnia thrived and their numbers quickly grew. Then one Friday a student brought in a live stickleback fish seined up from a nearby stream, wondering if I had interest in putting the two-inch-long fish on display. "Of course," I said, and popped him into the cladoceran tank. We spent more time that afternoon than now seems reasonable hunched over the sides of the glass captivated by the swimming cladocerans and the stickleback's robotlike actions gobbling them up one by one.

I should have expected what greeted me the following Monday morning when I found the tank empty of cladocerans. Not one remained. The stickleback hovered at mid-depth, finning himself in the center of the wide tank like a satisfied gourmand.

More than sticklebacks crave a cladoceran dinner. These crustaceans are a menu choice for the young of nearly every fish species. Rockbass, bluegills, sunfish, large- and smallmouth bass, crappies, and perch all gobble them up. In their first week of life, northern pike, walleyes, and muskies feed voraciously on them. As the young of these fish grow in size, feeding shifts to other foods, but without this baby food of cladocera, young game fish could not survive infancy. Cladocerans can make up as much as 90 percent of food volume in the stomachs of bait minnows.

Some fish, including tullibee (also known as cisco or lake herring) never outgrow their love of this baby food, as I discovered studying the species in northern Minnesota. Even in fish as large as a pound and a half, cladocera composed a significant portion of the identifiable contents of stomachs.

Okay, that pesky minnow is gone. You can relax—for now. Unfortunately for you, several other creatures, particularly the wormlike larvae of the phantom midge, *Chaoborus,* also relish cladocera. These insects are called phantom because they possess the ultimate in camouflage—near perfect transparency. About all one can see of chaoborids in a jar of water are the black, and certainly to a cladoceran, ominous eyes. Matched against the chaoborids' stealth of the invisible, speed of the sprinter, and the sure-grip grabber-catcher antennae, slower-moving cladocera have little chance of escape. Cladocerans are to lakes what mice and rabbits are to grassland and woods—food for other creatures in the system.

You daphnids have tricks of your own to foil potential predators. Phantom midge larvae emit a chemical that some daphnids detect that stimulates their bodies to grow a tail spine and enlarge the head into a pointed helmet. This trick makes the waterfleas too big to handle for many would-be predators. What a delightful sensory and messaging scheme must underlie this case of a body seemingly knowing just when such increase in size is needed, and responding. A comparable event in our own world would be if bodies of people newly resident in crime-infested neighborhoods spontaneously grew patches of bullet-proof skin on exposure to the random whine of bullets.

Many cladocera descend into darker, deeper waters by day, where they are more difficult for predators to see. They rise at dusk to munch in the algae-rich surface waters under the cover of darkness. *Chaoborus* also

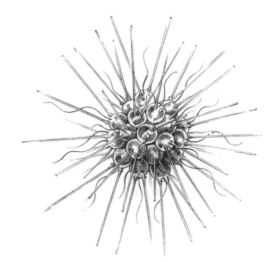

This green alga's spines make it difficult for a daphnid to eat this alga.

migrates in this fashion to save its own skin from predatory fishes. In shallow lakes like Diamond, cladocerans reportedly hide from predators by day in the cover of shore vegetation, then move out into the lake to feed at night.

· · ·

You must be getting hungry treading water all this time. Notice those many pairs of feathery leglike objects protruding from what you'd like to call your belly. No, they are not extra legs to help you sprint away from predators. They are eating tools. Flail them. Notice they not only create water currents but also filter algae, bacteria, and other tiny life and guide it to your mouth. Use your mandibles to grind things that need grinding.

Unfortunately, eating is not simple. Many algae have impressive defenses against daphnian predators. Some bristle with spines or have club-shaped bodies designed to make themselves too unwieldy for daphnians to eat. Other algae find protection in numbers, increasing size by joining cells together into colonies or long filaments. One even produces dozens of spines of glasslike silica, projecting in all directions from its colonies.

Imagine how discouraged you'd be about eating an apple laced with similar glass rods scaled up in size.

Larger size and bizarre shapes and structures do not always prevent algae from being eaten. But they do increase handling time required by would-be predators, reducing the overall number of algae that are eaten. Some algae thicken their cell walls, making themselves difficult for cladoceran mandibles to crunch. Others allow themselves to be swallowed but encase themselves in thick gelatinous sheaths, ensuring they can survive travel through the cladoceran gut.

One might expect Diamond Lake, with all its blue-green algae, would be a dining emporium for daphnids, an endless smorgasbord stretching to the horizon. You would be wrong. Blue-greens mount formidable defenses against daphnid predators in addition to forming colonies, gelatinous sheaths, and filaments. Some species produce obnoxious toxins that cause cladocera to reject them at the mouth, which explains why daphnids have not been able to filter this lake's water clear. "Smarter" animals, like dogs, wildlife, and cattle, can't detect the risk and, under conditions of high concentrations of blue-green toxins, may die from drinking the water.

The ultimate blue-green defense against cladocera is downright insulting. Some species sheathed in gelatinous envelopes not only survive the trip through the cladoceran intestine mostly unscathed, but on their way absorb nutrients from food being digested in the cladoceran's gut. These brazen blue-greens actually *feed* while being "fed on" and *increase* in numbers in the presence of daphnids. What in-your-face defiance of one's "predator."

· · ·

Some of my students discover even more exciting things than beating hearts. "Oh look! Mine has babies!" Their transparent coverings allow no secret pregnancies in the daphnian world. Sure enough. There, tucked into a brood chamber just below the heart, sit the youngsters like peas in a pod. Other students quickly crowd around for a closer look. What is the appeal of puppies and kittens and baby daphnians? What charmers they can be.

Nearly all daphnians are female. In some species males are unknown. Female cloning rules. Like mama like daughter. Making babies without sex actually has much to recommend it. No need to waste time and energy finding a mate or impressing the heart-throb in bizarre courtship rituals. No worrying about compromising one's own set of genes, one's own sex cells, with who knows what kinds of genetic mayhem some slick suitor might deliver. The random mixing of genes that results from sex produces a variety of traits in the babies that is sure to include a few genetic misfits unable to cope. In cold calculation, a waste of time and resources.

That the large daphnian in my jar has survived the challenges in this habitat to reach maturity reveals she holds a winning genome, has the right stuff, you might say, like the genetic equivalent of holding a four-of-a-kind hand, maybe even a flush, in poker. With no sex, no shuffling of genes to mess with, she simply hands off the flush, her superior DNA, to her babies, all of them. And babies with winning hands beget grandbabies with winning hands ad infinitum. With no sex involved, cladoceran populations can explode, quickly exploiting and dominating a favorable habitat. The cloning habit is widespread among other animal plankters for this very reason.

But habitat can change—suddenly—leaving Lady Daphnia's offspring in harm's way, holding a royal flush of gene cards but finding themselves no longer playing the ecological equivalent of poker, but perhaps hearts instead. A hand holding four queens, a likely winner in poker, can be disastrous in a game of hearts.

The daphnian genome has a wild card up its shell-like carapace. When the environment deteriorates, food grows scarce, or wastes build up from an overcrowded population, that card is played. Some of the unfertilized eggs that would normally become females become males. And, as night follows day, the sex that follows churns out variety in the next generation, improving chances that in the new circumstances a-coming, at least some of the babies might have the right genetic stuff to succeed. Sexual eggs can also go dormant, able to withstand freezing and drying and other environmental torments, biding their time until conditions improve.

·   ·   ·

I empty my jar and its waterfleas back into the lake, returning the captured life to its home. As class ends and cleanup begins, students commonly ask me what to do with the still lively creatures. "Do you want us to just dump them down the sink?" some plaintively ask, saddened in anticipation of my reply. "No, rinse them into the aquarium," I say, and watch tight lines of concern melt from my students' faces. Familiarity deepens relationships.

. . .

Water clarity is the preferred measure of lake quality for most people. The deeper into a lake one can see, the higher the price purchasers of lakefront property are willing to pay. One study of lakes in the Mississippi Headwaters basin revealed lake shore values rise or fall directly with water clarity. A three-foot reduction in depth of visibility reduces the value of shoreline by nearly $600 per running foot, a $120,000 drop in value for a 200-foot lot.

The connection to cladocera is direct. The fewer the algae the clearer the water, and hungry cladocera suck up algae like vacuum cleaners. In late spring and summer, when cladocera and other animal plankton populations are at their zenith, these algae eaters are known to filter the entire volume of a small lake—in just one day. Cladocera accounted for up to 80 percent of this filtration effort. Lakes without fish to eat cladocera had three times the visible depth compared to lakes with fish.

Who would ever imagine animals the size of seeds affecting property values or that the shadow of the lowly waterflea would fall on property tax statements and real estate transactions.

. . .

Delightful as waterfleas may be, their diminutive size and bashful daytime habits will prevent them from making direct impressions on casual lake watchers. Such is the fate of a plankter. But surely we can perceive them through the fish in the pan and the clear water of a bay. A tip of the hat to Ladies Daphnia. Small is beautiful indeed.

Limnos II—Fox Lake, Illinois

If one wishes to become acquainted with the black bass, for
example, he will learn little if he limits himself to that species. He
must evidently study also the species upon which it depends for its
existence, and the various conditions upon which these depend. He
must likewise study the species with which it comes in competition,
and the entire system of conditions affecting their prosperity; and
by the time he has studied all these sufficiently he will find that he
has run through the whole complicated mechanism of the aquatic
life of the locality, both animal and vegetable, of which his species
forms but a single element.

STEPHEN A. FORBES, 1887

The Chain of Lakes in the Fox River Basin of northeastern Illinois has attracted visitors since not long after the Civil War. The trickle of fishermen
and vacationers became a flood when the rail line from Chicago arrived in
1900. An advertising pamphlet published in 1909 carried ads for thirty-four
hotels and resorts, five taverns, four boat lines, two boat builders, and sun-
dry commercial establishments, though the resident population of the town
was a mere 400 souls. Today more than 3 million people recreate annually
on the area's waters.

Unlike these millions, Geri and I have come to Fox Lake not for recreation but to visit a very significant historical site. Though I expect there
are plenty of historic sites in northern Illinois, the one we seek does not
commemorate a military battle or the home of a great leader. I know of no
granite slab memorializing what happened here. That's an oversight. We
have come more as pilgrims to see where a perception was born and to
honor the man who changed forever our understanding of lakes: Stephen
A. Forbes.

At the moment we are puzzled. We cannot locate on our city map of Fox Lake and environs a public landing from which to launch our canoe. We drive north out of town looking for any road that breaks east in the general direction of the lake. After two false starts, we find one and turn onto it. The narrow ribbon of tar winds left then right, then up, then down, through a wooded residential area, fall leaves covering the ground. We find ourselves deposited in a cul-de-sac, stop the car, and look around for a way to get to the lake without trespassing. We have inadvertently blocked the way of someone backing a car out of a driveway. An elderly man walks toward us. I roll down my window. "Can I help?" he asks.

"We're looking for a place to get on Fox Lake," I reply. "Any public access ramps nearby?"

"No," he says. "It's all private property around here, but there's a park back at the edge of town where the highway goes over a bridge. That would put you on Lake Nippersink." He doesn't ask why we want to go on the lake on such a chilly windy day in November, so I'm not forced to try to explain.

We thank him and retrace our route back to the highway and locate a postage-stamp-sized Rotarian park at the water's edge across railroad tracks from the highway. A maroon truck, rusty around the edges, and a black van sit empty in the gravel parking lot, and two men stand on the shore fishing. One man is fishing for carp. The other, from Arizona, says he was told you could catch most anything here, muskies, northern pike, catfish, crappies, and sunfish, and he'd settle for any of them. Two lone individuals sit in separate cars looking out at the lake.

A sign announces no dogs, no booze, and the park closes at dusk. Another sign has a red slash through a red circle enclosing a stick figure swimmer. With the lake water so murky, the no swimming sign isn't needed. A sign at the cement boat ramp declares, "No Boat Launching. Strictly Enforced." I presume the restriction does not apply to nonmotorized canoes. No watercraft are in sight as we push off a gravelly shore into Nippersink and head toward Fox.

· · ·

Stephen Forbes launched an intensive study of these lakes over 120 years ago to better understand lake ecology. Influenced by the urgings of Louis

Agassiz at Harvard and others to raise the standards in natural history studies, Forbes had set out to examine in great detail the feeding relationships of
the creatures of lakes. How do populations of predators and prey, plants and
animals, remain as stable as they seemed, he wondered.

His essay "The Lake as a Microcosm," encapsulating insights gleaned
from his studies, became a foundation paper and launched him into the
pantheon of ecology's superstars. The essay describes his work on six lakes
of the Fox River Basin and Lake Geneva in Wisconsin. Since it devotes more
attention to his work on Fox Lake than to the others, that is the lake we've
come to see.

. . .

Shortly we round a point. A sign informs us we are now on Fox Lake. My
Secchi disc records a visible depth of just over nineteen inches. The water
has an unfamiliar gray cast to it, not the greenish color of algae I would
expect. Forbes notes that he pulled a plankton net across this part of Fox
Lake, and I toss out my net to do the same. I expect my net will capture some
of whatever gives the lake its color, but also, hopefully, some plankters.

Steel cribbing, cement slabs, large boulders, and planking protect the
shore, replacing whatever native plants once grew here. Houses and docks
and boat umbrella frames crowd the shore, cling to the shore, packed together like zebra mussels. In places, houses higher up the steep slopes
appear stacked on top of one another.

I retrieve my plankton net from the lake. The catch bottle is jam-packed
with a grayish material that looks silty, likely the same material that has
so greatly reduced the water's clarity. I lay the bottle on the bottom of the
canoe. I'll clean it out back at the landing.

We work our way up Lippincott Point. Forbes ran four sounding line
transects radiating out from this point and discovered the deepest part of
the lake lay at five fathoms in the middle of the west bay mouth. The great
density of homes and docks continues up the point.

We've seen only two boats traveling the lake, but the crowd of docks and
moored boats suggests things are different in the summer. A speed limit
sign prohibits travel faster than twenty-five miles per hour at night. What
a cacophony of boats, water skis, Jet Skis, and fishing lines must resound

on this lake on warm summer weekends. Is the apparent dearth of public landings intentional, intended as a screen to reduce boat numbers on the lake? How smothered with love this lake must feel.

We round the point and reenter Nippersink Lake into a strong head-wind. A man stands on the shore ahead of us. As we near he shouts at the top of his lungs to be heard above the wind, "YOU'RE NUTS!" He obviously questions the wisdom of being on the lake in this wind. We see nothing dangerous about this water. We have not taken a single wave over the ca-noe's bow. Hard work? Yes.

Back at the landing I fish a clear plastic cupcake container out of the trash barrel into which to drain the mud-like contents of my plankton net's collection jar. I release the stopcock, but the contents are too thick to flow out the rubber tube. I work the mass down with my fingers like toothpaste in a tube. The lower wad is now squeezed out and the rest, a thick fluid, flows into the container. Geri gets me a drinking-water jug from the car so I can dilute the material, stir it, and look for plankters.

I can't believe what I see! The "mud" is not mud at all, but a solid mass of waterfleas, a nearly pure culture of cladocerans glistening in the sun. My net sampled but a miniscule part of this lake. How many billions of them must be out there? What a thrashing the populations of green algae must be taking in this veritable sea of predators. Forbes, late in his essay, admits he saved for last creatures such as these, "minute crustaceans of a surprising number and variety, and of a beauty often truly exquisite." He devotes nearly a third of his essay to them and their relationships that tie the entire rich assemblage of diverse creatures of a lake, from bass to single-celled life, into a functional unit.

Where Thoreau and others had seen lake life as merely a collection of fascinating, individual life forms, Forbes saw something new. A lake isn't really, ultimately, about individuals, fish, or algae, or waterfleas, or water plants, he wrote. From his studies on Fox Lake and the others, Forbes saw through the chaos to the organic relationships that intimately connect all the living things in this lake into a community. In seeing connectedness and unity in chaos, Forbes altered for all time our perceptual understanding of a lake. Perceptual leaps may not be as flashy as military victories or the libraries of past presidents, but sometimes they matter more.

• • •

One might logically expect one perceptual advance to facilitate another. Paradoxically, in Stephen Forbes's case, just the opposite happened. The ironic twist occurs in the second paragraph of "The Lake as a Microcosm," first presented at a gathering of the Peoria Scientific Association on February 25, 1887. Forbes writes, "The animals of such a body of water are, as a whole, remarkably isolated—closely related among themselves in all their interests, but so far independent of the land about them that if every terrestrial animal were suddenly annihilated it would doubtless be long before the general multitude of the inhabitants of the lake would feel the effects of this event in any important way." With such strong language about a lake community's independence from the land, Forbes unknowingly delayed for decades another critically important perceptual leap forward in the understanding of lakes. But then, even Abraham Lincoln made mistakes.

Discovering Eden

Most have not delved six feet beneath the surface, nor leaped
as many above it. We know not where we are.

HENRY DAVID THOREAU

Simple observation, unhurried and deliberative, reveals much about life on the beach but little about life beneath the waves. Floating leaves, submerged vegetation, compromised water clarity, water depth, and reflections off a lake surface often obscure the underwater world. Much as peering into a forest from a helicopter hovering above the treetops, the constrained view from a boat or end of a dock gives scant account of the mostly unseen world below.

My curiosity about this invisible world surfaced when, as a kid, I discovered I could take a deep breath then dive and cruise the bottom with eyes open wide. Scurrying crayfish, clam tracks, water insects, and other unfamiliar sights fascinated me. Between visits to lakes I even practiced holding my breath to lengthen the duration of my dives.

At first I obeyed Mother and the beach rules and stayed within designated swimming areas. Later I discovered such sandy places are wastelands of emptiness compared to what lurked out of bounds. The truly fascinating sights, all manner of strange creatures and tiny fish, inhabited edges of weed beds beyond buoys and roped-off zones.

A sabbatical trip to Florida in my early years of teaching reintroduced me to underwater haunts and to prescription goggles. Watching a large crab on the sea floor while snorkeling in the Keys, I wondered, why can't I do this at home? Water too cool? Lakes too murky? Expectations that only coral reefs are worth seeing?

I have come to this small clear lake in east-central Minnesota to snorkel in a body of freshwater for the first time. I paddle my canoe past the swimming beach, along a wooded shore, and tie up to a birch tree that has fallen across cattails into the lake. With mask, snorkel tube, and fins, I flop onto my belly in the shallow water and wiggle and flipper and belly-scrape my way into the lake.

I submerge and look out among scattered plants. Poof—out of nowhere—as though sprung from a magician's hat, six tiny sunfish appear. They move toward me, as though curious as to my intentions. More fish appear, blue gills and pumpkinseeds of all sizes, several dozen in all. The gathered throng remains politely motionless, as though expecting me to call the meeting to order, or begin preaching, or deliver long-awaited news from the world above. Several fish, larger than the rest, hold back behind the crowd, eyeing me more warily than others. I turn and—zip—all vanish into the grove of plant stems that surrounds us.

I swish my fins slowly, leave the clump of plants, and head down the shore where a tangled mass of grayish-green stems blankets the bottom as far as I can see. The plants feel springy and crusty, like touching loosened steel wool, or blueberry brush. I pull up a handful, surface, lift my mask, and take a sniff. (Quirky people, biologists. We've an almost unconscious habit of checking new discoveries with our noses.) It smells like skunk. Identity revealed. It is the *Chara* I first met in aquatic plants class. Technically, it's not a plant at all but a plant-sized green alga. Its crustiness comes from a thin layer of calcareous marl that accumulates over its surface. I dive to look more closely. A school of minnows scatters in front of me, melting into the minute labyrinthine spaces of the finely branched *Chara*. What fantastic hiding places for young fish and the microscopic life on which they feed. Much of the lake's bottom just offshore is a vast *Chara* meadow.

Some lake cabin owners do not appreciate *Chara*, preferring sand beneath their toes to softly crunching algae. Yet *Chara* keeps waves from riling bottom mud and dirtying the water. Its tiny spaces provide sanctuary for waterfleas that sally forth to eat algae that would otherwise make the lake murky green.

Ahead of me a group of spindly, stringlike stems spiral lazily to the surface. Each stem ends in a dainty white flower the width of a pencil eraser, so small and inconspicuous, from a boat one might not even realize they are flowers. Wild celery. *Vallisneria.*

I also notice tiny white floating specks no larger than the diameter of pencil lead. I recognize them as male *Vallisneria* flowers, floating at the whims of the winds perchance to encounter female flowers of their species. Few sexual strategies are as delightfully peculiar as those of this plant. Male flowers grow at depth. When mature, each is sealed with a bubble of air in a

Chara stabilizes bottom sediments, keeping lakes clearer.

special encasement and released. The buoyant bubble floats to the surface, where the capsule bursts open, releasing the male flower to drift about the lake. Female flowers create a dip in the water's surface film, producing a slope, a miniaturized version of a children's playground slide, that leads to themselves. When a male flower sails close enough to the sloped water, it glides down the slide into the embrace of its mate. How romantic.

Ducks and swans and other wildlife love *Vallisneria* and eat every part of the plant. Canvasback ducks are known to change flight routes to seek it out. The scientific name for the canvasback, *Aythya vallisneria,* acknowledges its fondness. Wild celery also attracts marsh and shore birds and muskrats and makes good fish habitat.

My fins propel me toward another group of plants and more fish appear. As one group swims off, another replaces it. Several dozen at a time are not uncommon. Not all are sunfish. Often, a lone largemouth bass appears, or an occasional small perch hovers at the edge of a sunfish school, then wanders away, obviously less curious about me than the sunfish. Oh, the lake life I have missed from the canoe!

Streams of incredibly tiny bubbles, each the size of the head of a pin, emerge from the tips of several plants. Like so many dainty fairy necklaces, these glistening strings of pearls ascend radiantly, sinuously, toward the surface. I reach to touch one and the serpentine string sways gently away.

A large patch of lilies with their white and yellow flowers grows at the west end of the lake. The idea of snorkeling through a thick stand of lilies, foot-wide leaves forming a canopy over the water below, seems odd, I suppose. But how can one learn what the patch is like without going in?

Entering the edge of a lily colony is much like walking into shaded woods. Sunlight striking open water becomes muted shade the moment I part two green stems with my hands and enter the lily grove. The water is cooler as well.

Fish in small schools arrive immediately, as lily stems sway in gentle rhythm with waves at the surface. The mass of plants is not as impenetrable as I had first feared. Guiding myself with outstretched arms, I travel a slow-motion slalom path, turning this way and that to stay within the more open leads through the stem curtains. Both leaf and flower stalks are much longer than needed to reach surface from bottom. This slack makes it easier to push stems aside. It also gives plants the flexibility to safely ride out angry waves that might otherwise rip leaves from stems and stems from roots.

I take a deep breath and dive for the bottom, intending to remain there as long as my air holds out. But bodies prefer to float. No sooner do I reach the lily rhizomes than, against my wishes, I begin floating to the surface. The remedy, I discover, absent a weight belt, is to grab hold of a lily rhizome. I dive again, find one as thick as my arm, and hold on for dear life.

Now anchored, I look up at the surface. Backlit by bright sun, the ceiling lights up into a hundred shades of green. Small pools of water sitting on the upper surface of the leaves appear as green splotches, contrasting with the lighter colored leaves, dark green clouds floating in a grass green sky. Other leaves, in stages of decay, add splotches of amber and yellow and brown.

Random beams of light penetrate the ceiling unhindered through gaps

Vallisneria is among the most important of all waterfowl foods.

between leaves and, like a dozen tiny spotlights, draw my eye from one highlighted object to another. A young lily leaf, the fresh green of new life yet unfurled. Brown diamond-shaped rhizome scales highlighted against the glistening white of the rhizome body.

A particularly large beam spotlights in brilliant radiance the pièce de résistance. Bathed in the light of center stage, the tender green bud of an unopened lily flower emerges from the tip of an uncoiling stem spiraling out of the bottom's darkness. Two narrow cracks in the bud's green envelope emit brilliant slivers of pure white, giving promise of the burst of glory to come when the bud reaches the surface.

No wonder magical powers have been attributed to this plant since medieval times. An ingredient of love potions, the flower must be picked under a full moon, and those who collect it must plug their ears to avoid becoming bewitched by water nymphs. Bewitched indeed. But a body's scream for oxygen cannot be resisted indefinitely. I hold onto the rhizome to the last second then release my grip and zoom to the surface through a forest of stems.

One sets aside all sense of personal dignity venturing into such places. Emerging with a lily leaf the size of a dinner plate flopped sideways over your head and draped over an ear, swim mask and snorkel tube hopelessly entangled in green stems, quickly sweeps respectability away.

· · ·

Countless stems populate shallow waters, and a layer of tannish green, I could call it *gunk,* covers them all. On some plants I touch it is thick like fuzz. Others feel slimy or even fluffy. I've also seen it black or brown. None elicited thoughts of beauty. Aquarium keepers know that left unattended, aquarium walls become less clear by the day, as gunk grows from a smudge into a yellow-green-brown layer so thick you can't see fish through the glass. Gunk to some, it's known as periphyton to others (*Peri* means around and *phytom* means plant in Greek).

No need to snorkel to see it. On rocks and docks, plant stems and leaves, and sunken tree branches—it's nearly everywhere in lakes. Mundane and ubiquitous, periphyton is easily overlooked as unimportant. In truth, periphyton is a rich amalgamation of bacteria, algae, fungi, single-celled creatures that creep or swim, myriad tiny animals, and sometimes marl, mucilage, silt, and tiny bits of debris from decaying plants.

Aquatic plants gain no benefit from this blanket on their surfaces. Periphyton sucks up nutrients that would otherwise be available to the host plant, and in some cases obtains nutrients directly from the plant itself. It also screens out as much as 80 percent of the light that would benefit the plant.

Some plants and large algae produce transparent mucilaginous coverings to prevent periphyton from taking hold. I run my fingers along a pondweed leaf. The periphyton layer feels thinner than in other lakes. Nutrient-poor waters can cause that, but in this case, I think it is snails. Windrows of empty shells pave the shore, and live snails cling to plant bodies everywhere. In places they are so numerous I can't avoid crunching the shells with my step.

Many animals make a living grazing on periphyton. This growth turns even the most austere rock surfaces into bountiful gardens of things to eat for creatures from mayfly nymphs to amphipods, and crayfish to snails, especially snails.

That periphyton attracts snails and other eaters is no surprise. It contains twice the nutrients of other aquatic foods and is more nutritious than the plants themselves. What's more, it can neither run away nor bite back. Snails and other grazers can remove half the periphyton crop produced in some lakes each day. By eating periphyton, snails keep plant surfaces and aquarium walls clean. When pumpkinseed sunfish feed heavily on snails, periphyton grows lustily and plant growth declines.

. . .

Two pumpkinseeds hover off to one side and below me. One suddenly darts to the bottom, snaps up a snail three-quarters of an inch long, then promptly spits it back to the bottom. These fish are specialized by jaw anatomy to prey on snails. The attacked snail was apparently too large for the fish to handle.

Snails seek cover when sunfish are present. That likely explains why I see so few small ones. Sunfish have a window of opportunity to eat snails when they are small, provided they can find them.

A chill sweeps over me, and I stand, shivering, to get my bearings. A pair of swans with three cygnets float between a lily patch and the shore. One tips herself upside down, submerging neck and head to form a perfectly perpendicular line with the water, tail end pointing skyward the way an Olympic diver in top form appears at the moment of water entry. Now she rights herself and spits out a mouthful of plants onto the water. Cygnets scramble to gulp it down. Over and over again, she and her mate repeat the sequence as insatiable youngsters splash and lunge for the obviously delicious fare, periphyton and all.

I return to the canoe, dry off, and break out lunch, my mind reveling in the wonders of the morning as taste buds savor food. Refreshed, I paddle down the lake to resume snorkeling off a sandy shore. I launch myself and wriggle directly into a dense stand of plants with threadlike stems and leaves, *Najas* (nigh-us), named for the water nymphs in Roman mythology that give life to a lake.

Now over deeper water, columns of green plants the shape of long bushy squirrel tails appear, soft and pliable to the touch. The bushiness comes from whorls of small leaf stalks, each stalk bearing many pairs of short threadlike leaflets resembling the structure of feathers. This is milfoil, abundant and native to North America (not the invasive Eurasian species). Its lacey fanlike "leaves," a leaf type common among plants committed to life submerged, hardly seem like leaves at all. A plant with such leaves would quickly dehydrate and die on land.

A turtle emerges into an opening in the milfoil three feet down, turns, paddles slowly toward me, stops, eyes me intently, now disappears behind plants. Within his camouflaged refuge, he moves stealthily to a narrow opening in the plants and pokes the tip of his head through the slit to spy on me once more. Pessimism carries the day. He turns suddenly, thrashes his legs, and disappears into a jungle of green, knocking loose a shower of gray marl that settles slowly to the bottom.

Plants thin out in deeper water and become taller. Milfoil cylinders reach up from the dark bottom heavenward, like skinny church steeples without the churches. I now enter a gathering of skinny stemmed aquatic plants of a group called "pondweeds." "Weed." What an unfortunate and misleading name for these plants. Language matters. Before being seen, before revealing anything about their lives and relationships, they stand condemned. Useless. Nuisance. Undesirable. "Pondweed" is not simply a generic name for aquatic plants, despite its common use that way. Pondweed is also the name of a large and grand family of aquatic plants known more technically as the Potamogetons (Po-ta-mo-gee'-tons) (from Greek, *Potamos,* river and *geiton,* neighbor). How different our perception of these plants might be had we retained the Greek root and called it "pond neighbor." What power the namers-of-things can have over attitudes. (I must note the invasive exotic pondweed species, curly-leaf, so disrupts our lakes it richly deserves to be called a weed with all the negative connotations.)

Again, as before, when I near vegetation, fish come, schools and loners, fish the length of my little finger, fish the size of my hand. One sunny, not three inches long, swims right up to my mask and we stare at each other, eyeball to eyeball. What on earth is he thinking? I could swallow him whole in one gulp. What drives his curiosity? Maybe he is inspecting me for periphyton, judging whether it's worth his time to look me over for snails, except his gaze is transfixed on my face. Were he to do this to a largemouth bass, he'd be swallowed in an instant. In a few moments he turns and swims lazily away. What has turned off his predator-awareness sense? Has he determined my movements are too clumsy to pose a threat to him, no more dangerous than a submerged log? Two sunfish nip at my ankle and another bites at the shiny metal of my wedding band. I can hear the click of mouth against metal.

A largemouth bass glides into view and mingles with the sunfish. These bass are sunfishes' main predator. Why don't the sunnies flee in fear into the surrounding potamogeton forest? They must sense he is not large enough to eat them.

This bass keeps his distance. A skilled predator, he understands the concept of strike-distance as well as I do. No predator can afford to cruise around wasting energy striking out at any prey fish in sight. Bass instinctively understand that strikes are most often successful when the intended victim is no farther than a foot and a half away. Bass that follow the foot-and-a-half rule are deadly, nailing their prey 70 to 80 percent of the time. Is that what's going through this fish's mind as we look each other over? Is he trying to determine what *my* strike distance might be? He swims off into the plants.

As the water deepens I hyperventilate for greater air reserve and dive into an open space among scattered potamogetons. I spot the largest bass I've yet encountered. At nearly three pounds, what a catch he'd make. When he appears to lose interest, I swim slowly away then look back. He has followed me. I swim farther and look back again. He's still there. I maneuver between two large clasping-leaf pondweeds. There he is, still behind and below me. Is he stalking me? Surely he can't intend to attack. Though when bass pursue prey, the first step is to follow the potential victim. Is his behavior toward me the compulsive instinct of the predator that cannot be resisted, like dogs chasing cars, ancient yearnings too deeply embedded to deny, even when the pursued is not vulnerable? Maybe he likes the cover I provide for him. I once watched a large sunfish hover beneath my canoe, moving to remain in its shade. Or maybe he's noticed how sunfish attracted to me are the right size for him to eat and he's using me as a decoy. I feel strangely uneasy.

I see other potamogeton species. One lies on the bottom with leaves in a single plane resembling a fern. Three others have leaves tightly clasping the stem, giving these plants a cylindrical look.

Now a different pondweed. This one's broad banana-peel leaves droop like the floppy ears of a lovable rabbit. *Potamogeton amplifolius,* broad-leaf pondweed (*Ampli,* large; *folius,* leaves). A massive stand of broad-leaf appears ahead. The droopy leaves transform the lacy world of milfoils and other petite leaf plants into one of coarse texture, a place of overhanging balconies, of places to rest and hide and ambush. I understand immediately why it came to be called "bass weed" and "muskie weed" by fishermen. This plant, together with the rest of the potamogeton tribe, is the largest source of food for waterfowl. It also harbors food for fish. Broad-leaf is picky about water quality, and is one of the first plants to disappear from a lake when quality declines. That such large colonies thrive here deeply satisfies. The water nymphs tend their garden with loving care.

The bottom slopes sharply away to greater depth, and plants become less crowded. I reach the outer edge of vegetation and look into deep clear water. Long spindly green fingers of clasping-leaf pondweed reach for the surface in a magical world where the laws of physics seem no longer to apply. The plants grow heavenward like unassisted vines, as though gravity has no claim on them. Scattered individuals of the small-leaf species, more spindly yet, suspend themselves motionless. But I perceive with the terrestrial's mind. Water's buoyancy and endless air chambers embedded everywhere within these frail bodies only make it seem that gravity sleeps.

Several sunfish schools coalesce and swim down an alley between plants. I follow, like the last float in their parade. They head out over deeper water, so I stop. And, of all things, the fishes stop too, and most of the school swims back toward me, as though wondering what's the matter? Why have I stopped? They act concerned that I'm okay. Why do they pay any mind to my leaving them? By swimming benignly among them have I unintentionally entered membership in a finny brotherhood? Their attention moves me.

I watch them watch me, then turn to resume my travel paralleling the shore. I stop suddenly and look back, and there they are, following me. I feel like the Pied Piper. I quick-stop again and they nearly bump into me. Underwater slapstick. A Three Stooges routine.

· · ·

Now periphyton smothers surfaces in a filmy yellow-green shroud. The leaves of the pondweeds become scaffolding for algal filaments to bridge open spaces. I dive. Periphyton blankets the lake bed with a gently undulating layer of tan gauze the texture of cotton candy. Periphyton triumphant. All spring and summer the plants, with the help of grazing snails, have grown faster than periphyton. Now, in late summer, the tide has turned. Many plants are slipping into senescence. With plant growth all but halted, periphyton has finally overcome. Senescent plants become decaying plants, then nonexistent plants, and the living creatures of the periphyton, without substrate, must die and decay as well.

I collect pieces of milfoil, fern pondweed, and water crowfoot and swim to shore to photograph them. As I submerge them in water in a white pan and straighten them out, the water comes alive. Tiny waterfleas, damselfly larvae, seed shrimp, even planaria skitter and slide across the pan to new hiding places. Every one of them a tasty morsel for a minnow, a sunfish, a bass, even tiny walleyes and pike. I have flushed the multitudes from their homes.

· · ·

I reach for dry clothes and look at the world I am leaving. Lake water looks to my eyes exactly as it did long ago, but not to my mind. My mind frolics, enveloped by this soft and delicious place of intimacy and shadow, mystery and illusion. Where plants grow as though their architect didn't believe in gravity, where fish float as if on ethers, like birds riding invisible thermals.

My brain revels in these new exotic dispatches from the eyes. It's been said "vision" is comprised of one part input from the eyes and four parts from the brain. Like the sunfish peering through my mask, unsure what to make of this strange intruder, my mind now blends the new with its own sense of things.

Swimming among the potamogetons is incomparably more inspiring than sorting out their anatomical differences on a lab table. Their world is vastly different, vastly richer and more interwoven than any I could learn about looking over the side of a canoe. I have experienced them in *their* place on this earth, and that has made all the difference.

Can one come to love the aquatic plants of a lakeshore, those despicable weeds in the minds of some? Is love a spontaneous, altruistic feeling of affection toward something that has touched one's emotions? Though it be homely, festooned with periphyton, sometimes crowded but always sublime, having experienced this enchanting place, how can I feel otherwise?

Does it at all matter? Essayist Scott Russell Sanders says, "We treat with care what we love, and we love only what we have truly learned to see, with all our senses alert."

Most people will never enter this enchanted world with snorkel tube and so will never experience the transformation that turns the mundane into magic. Can virtual perception stand in for experiential perception? Oh, how I wish that it could. The vegetated zone of a lake is more than a place to delight the senses. It is a sacred garden. Loss of the garden plucks the pulsing green heart from the lake.

· · ·

A loon calls a single note from midlake, a coda in the song of the garden, a sign my adventures have ended for the year. Loons will soon flock up and fly south. Thoreau played a game with the loons of Walden Pond, guessing where they would pop to the surface next. He always lost. Thoreau attributed metaphorical powers to loons. Because they could penetrate the surface and dive deeply, they could get closer to the hard bottom of truth than humans. "Seek truth?" asks the loon. "Come beneath the water and find it."

Lake Agassiz's Child

Landscape is not "land," it is not "nature," and it is not space.…
A place owes its character to the experience it affords to those who
spend time there, to the sights and sounds, and indeed smells that
constitute its specific ambience. And these in turn, depend on the
kinds of activities in which its inhabitants engage.

TIM INGOLD

Lake Winnipeg lies in the Canadian Province of Manitoba, some seventy miles north of North Dakota and Minnesota. It is the largest surviving remnant of Glacial Lake Agassiz. Geri and I left the Winnipeg suburbs several hours ago heading north to see what we can learn from this, the tenth largest body of freshwater in the world.

The ends of our double cockpit sea kayak stick far beyond the ends of our car. The boat looks like a streamlined red-and-white pterodactyl with folded wings that could take flight on a whim, carrying the car off as though captured prey.

We arrive at Grand Rapids, situated on the northwest shore of Lake Winnipeg, Mistehay Sakahegan ("Great Lake") to the Cree, where the Saskatchewan River empties the waters of a thousand prairie streams into the big lake. We arrange to leave our car in the police parking lot for the sixteen days we expect to be on the lake and set up camp in a primitive public campground at the edge of town.

Morning awakens calm and sunny, the midnight thundershower now far east across the vast lake. Two pelicans sit on floating wood as we launch at the mouth of the river. Arms and paddle soon flow in relaxed rhythm. At 260 miles long and 60 miles across at its widest, this lake is a third the size of Lake Superior. We plan to paddle north along the west shore, then east across the top of the lake to Warren's Landing, then on to the Cree settlement at Norway House. We will return to our car by the same route in reverse, 230 miles in all.

. . .

The lake has a legendary reputation. Numerous lake freighters and lives have been lost to its storms. The lake has been described as a "boiling, raging cauldron of unleashed fury." Experienced wilderness paddler Roger Terenne says, "It happens so quickly. Lake Superior gives you fifteen minutes from calm to dangerous waters. But with Lake Winnipeg you get five." Adventurer Phil Manaigne has said, "It's definitely the roughest meanest lake in Canada. One minute it charms you and fifteen minutes later it tries to kill you."

I understand. I paddled this lake with college friends decades ago, before I met Geri or read those words.

On the map the shoreline subtends a broad arc for over ninety miles, west to east, across the lake's north end. Smooth curves on a lake map often mean sandy shores. The beaches here are of limestone pebbles, limestone blocks the size of children's tables, and limestone sea cliffs.

Modest wind and waves and Geri's aching forearm slow progress, and we decide to land on a steeply sloped pebble beach. I fear for the kayak's thin skin in an encounter with the sharp limestone. We leap into the water and finally manage to drag the kayak over driftwood onto the beach. A limestone block provides us a perfect kitchen, and we discover a tent site among the large stones.

Morning. The lake is glass. Half an hour into paddling a wolf howl breaks the stillness. Now we see him, coal black body sauntering down the long pebbly strand headed in our direction. Our paddles freeze. The craft's momentum carries us silently across the shining surface on a line toward shore. The distance between wolf and kayak closes and the animal, incredibly, doesn't sense our presence. We watch in breathtaking silence as he trots leisurely past. How magnificent!

A limestone cobble island glistens white with pelicans, gulls, and terns. They rise in an alabaster cloud and drift out over the lake as though the land itself has levitated and floated away. Only the cacophony of bird voices reveals truth.

What a magnificent bird, the white pelican. White body, black-fringed wings spanning up to nine feet, ridiculously large bill and mouth pouch—a master of graceful flight. I've watched pelicans rise on thermals, becoming mere specks in the sky, visible only when flying crossways to me, disappearing as the flock turned on the spiral toward me. Now you see them, now you don't. The magic comes when they defy gravity and glide mere inches above the water, wingtip to wingtip, in motionless motion.

A pine marten in glistening black fur scampers across limestone pebbles from the water's edge into the woods. A relay of eagles flying shuttle escort us down the shore past beautiful sea cliffs. I see little sand. Limestone lacks the tough minerals, the quartz and feldspars, needed to make sand. The spicy aroma of balm of Gilead floats over the water as limestone finally gives way to sand and we camp. Driftwood of whitened beaver sticks, weathered knots, and bone dry limbs spring to flame with the match. The wood pops and crackles, and the smell of smoke wafts from our cookfire.

Mealtime is surprisingly free of mosquitoes. The many spiders and their curtain of webs at the edge of the woods must take a heavy toll. This lake is utopia for spiders. A feast of 90,000 metric tons of mayflies are estimated to arise from its fertile waters each year.

We cross the three-mile-wide mouth of Limestone Bay to the western end of Limestone Point in quiet water. Over the years the southeast wind, with a 150-mile head of steam, has massively moved sand from east to west to create this twelve-mile-long and growing peninsula. Two parallel sand ridges on the upper beach create a roller-coaster beach profile. Pushed by powerful spring winds, lake ice has creaked and groaned up this beach many times, pushing sand, camp fire rings, and all else in its path, back into the trees.

Five days out we reach the base of Limestone Point. The beach narrows dramatically, sand replaced by high, eroding clay banks eating away at a spruce bog perched above, creating a vertical bank fifteen feet high. A jumble of fallen trees, roots eroded out from under them, and blocks of clay and coal-black peat have slid to within a few feet of the water's edge. The lake is washing away the land faster than its waves can create beach. Much of the erosion here results from hydro development of Lake Winnipeg's waters as they flow into the Nelson River and ultimately into Hudson Bay. Hydro development, in addition to causing great distress for the native Cree, has also turned Lake Winnipeg into a giant reservoir.

A mother bear and cubs scramble up the steeply pitched bank and disappear into the bog forest above. For long stretches, the clay beach is so narrow we dare not set up camp. Even a slight rise in the wind would send waves crashing into our tent. We reach the mouth of a wide artificial channel, dug as a second outlet from the lake to increase water flow to hydro-dams on the Nelson River downstream, and take this two-mile canal as a shortcut into Playgreen Lake and on to Warrens Landing, where the lake narrows and forms several channels before eventually forming the Nelson River.

Several miles from the landing feelings of nostalgia come over me. The Lake Winnipeg leg of our youthful canoe journey to Hudson Bay ended at this remote outpost. The landing serviced lake freighters that off-loaded cargo and passengers from the south to smaller boats to continue on to the Indian settlement at the historic fur post at Norway House.

We finally arrive at the landing. A fringe of brush lines the shore. I stare in disbelief. Where are the wharves with lake freighters tied up alongside? Where are the fish boxes, aviation fuel drums, sealed crates, and the roustabouts who should be carrying such things in and out of the big boats? The Indian kids? The mess hall? The string of tiny barn-red cabins where fishermen stay? Whitefish boats? Where are the people? What has happened to the place I once knew? All turned to bushes and a grassy field.

Down shore I spot a small dock with a boat and outboard motor tied alongside. Voices drift out from behind the overgrown brush. We land and join a small group of people standing nearby. They are having a reunion. One of the men, George Teasdal, is from Selkirk. He captained a lake freighter years ago. "Used to haul logs from here to the Winnipeg River," he says. "But once the road came through to Norway House, it was over. The lakers are no more. Most everything travels by land or air now."

I explain how I came to know this place. "Where are the whitefish boats?" I ask. "Commercial fishing is not like it was thirty-five years ago," comes the reply. Whitefish, along with walleyes, are still the mainstay of the commercial catch. A quota entitlement system established in 1985 was intended to reduce the number of fishermen and thereby increase the incomes of those that remained. When quotas are filled, fishing stops. The goldeye and sturgeon fisheries have not yet recovered from population collapses caused by overfishing decades ago.

Pollution has increased significantly since my earlier trip, and some feel it threatens the lake's ability to purify itself. Phosphorus input from the lake's immense watershed is equivalent to concentrations that formerly entered Lake Erie before preventive measures began. Lakes with shallow basins and large watersheds are most vulnerable to degradation by human activities. Lake Winnipeg, with an average depth of only twenty feet and immense watershed, nearly 400,000 square miles, second largest in North America, make it particularly vulnerable to human perturbations. Lake Winnipeg is fortunate in one respect. The lake turns over, replacing its volume, more quickly than many other lakes, carrying pollutants away, allowing people who live on its watershed in Alberta, Manitoba, Minnesota, North Dakota, Ontario, and Saskatchewan to modify land use habits to reduce impact on the lake. Ecological transgressions are not as easily undone elsewhere. Lake Superior measures its water replacement time in centuries.

Global warming models project significant increases in lake water temperature. Such warmth, depending on its magnitude, could potentially disrupt Lake Winnipeg's valuable whitefish fishery. That species relies on cool-water plankton for food. Mercury loading has increased markedly since I first paddled the lake.

I ask about the *Lu-berg,* the lake freighter that brought us boys south to Selkirk when we returned to Warren's Landing from the Bay. A smile lights up George's face. "You should have been here this morning. We had a gathering of old boat captains to reminisce about the old days. The *Lu-berg*? Let me think. She might be one of those they've put in that boat museum at Selkirk. If she's not there, she's probably rotting away in the Selkirk slough."

Several small neat buildings sit in a clearing back from shore. "A few families have summer cabins here. Otherwise, the landing's abandoned," a man says. How sad, I think.

The people board their motorboat to return to Norway House, the road head, and the present. We wave each other good-bye as they fade into the distance. I look back wistfully at the emptiness. The passage of time has plucked from me a part of myself. We turn our backs, return slowly to the kayak, and paddle onto Playgreen in search of a campsite.

Playgreen Lake drops a full foot overnight. Apparently the hydro people downstream need more flow through the turbines. Minneapolis air conditioners must be working overtime.

. . .

The big lake is rough when we return from Norway House to the channel entrance. We are windbound. Wind delay can frustrate, can push one into launching when one should remain on shore. It also frees time to nose around, to think and reflect and watch. We have extra food and our car is in no hurry.

Instinctively, water wants to lie undimpled, smooth as polished glass. A river of air flowing over a water surface delivers some of its energy to the water, inciting a rebellion against gravity's tyranny. First come ripples. As air becomes more energized, ripples become waves. Waves are deceptive. Much of their story is hidden from the eye, and the truth about them defies common experience.

I've walked into large waves on a beach and felt them push me backward, even knock me down. I've also treaded water among similar waves out from shore. As waves come along one by one, they bob me up and down with each crest and trough with hardly a nudge toward shore. The waves themselves continue landward, leaving me behind, quite alone. Why should the same waves treat me so differently on the beach? My eyes are convinced a wave is a specific volume of water racing to land. The truth is more intriguing. Instead of pushing water horizontally across the lake, much of the wind's energy creates vertical elliptical water currents that move like a rhythmic elevator up and down. These currents raise the water up creating the crest of a wave. Just as a bug perched on a jump rope that is whipped into up and down waves is not moved toward either end of the rope, neither am I, the swimmer, moved toward shore. Both bug and swimmer merely bob up and down.

When energized to exuberance, these currents produce waves frothing with whitecaps now dancing across the lake toward me. What once was a crest quickly drops into a trough as gravity brings the rebellious water back under control. I watch a large wave approach land. When the wave reaches shallow water, its contact with the lake bottom disrupts it, transferring its energy into forward motion, knocking me down and pushing sand and flotsam farther up the beach.

We laze around this pelican loafing ground, pick raspberries, cook a supper meal for lunch, and keep an eye on the big lake. In late afternoon Geri detects a wind shift. It now blows off shore from the north. What luck! We load the kayak and paddle west along the clay battlements. Mile after mile we hug the land, steep clay banks to our right, a hundred fifty miles of open water to our left.

Muscles tire and we search for a place to camp. We find none. The beach is simply not wide enough to squeeze a tent between lapping waves and the base of the tumbled clay bank. We paddle on. Well into evening we find a wider spot of beach and pull in. Only our muscles are pleased. We pitch our tent as far from the lake as we can, to the very edge of the jumbled trees and peat, yet we are only a few feet from the edge of the water. We are under no illusions. Cramped between the clay bank and the lake's edge, all will be fine only as long as the wind doesn't change direction again. It will not take much wind from any southerly direction to inundate the tent. Waves from a strong wind will simply wash us away.

Geri is uneasy, but we have a contingency plan. Should conditions go sour, we will drag our tent and packs up onto the jumbled slope. Though the tent will be unpitchable we will use it poleless, as a shroud, and rest in whatever positions the tangled trunks and blocks of peat allow.

We drag the kayak as far up the steep slope as we can and rope it securely to a spruce trunk that has recently toppled head first down the bank. After rye crisp and jam and butterscotch pudding, we crawl into sleeping bags—and listen. Do the tent flaps stir? Is the wave sound changing? Unlike Geri and I, the lake sleeps through the night.

Morning brings, of all things, glassy water. The lake is sleeping in! We gulp down energy bars, rush to break camp, and paddle. In hours we reach the flat ground of Limestone Point. But clouds in the west herald an approaching front, and the inevitable waves begin building. The lake now kicks up quite a fuss. Waves, intimate and soothing, rock a baby to sleep. Lapping gently against a shore they calm the troubled soul. Angry and crashing, they sober and frighten and can take a person's life.

Windbound. Reality on this lake never changes. I finish reading my book, *Company of Adventurers,* a history of the Hudson Bay Company, and collect seed pods from a beach pea. Some pods are dry and mature, while others remain in purple flower. Geri looks for driftwood. We find tracks. Too large for marten, might it be a young wolf?

A new day and the waves still pound. Is this lake a predator in disguise, ever eager to pounce? We consider launching but decide the risk is too great. This damnably endless wind wears you down. We carry the canoe far up on the beach.

Lake Winnipeg is shaped like a fat exclamation point. A stretch of water only a few miles wide separates the larger north basin from the smaller "dot" basin to the south. The long axis of the lake tilts northwest to southeast, precisely the directions from which the prevailing winds blow. That quirk of geography produces the longest possible fetch for those winds, delivering the maximum energy and power to waves. Unfortunately for lake travelers, Lake Winnipeg lies in the belt of strong prairie winds. The lake's long fetch means wind speeds that could never stir mud off a ten-foot-deep bottom on a smaller lake can resuspend mud from many times deeper on Lake Winnipeg. Turbulence touching bottom creates murky, silty, water, and resuspends phosphorus from the bottom muds, increasing algal and cyanobacteria growth.

I have learned that "persons" in the Ojibwa tradition can exist in many forms, of which human is just one. Not only can animals take on other-than-human personhood; so too can heavenly bodies, stones, and even meteorological phenomena. Such persons are known as "grandfathers." Grandfathers are more powerful than humans, in that they can change form. The Four Winds were among the most powerful grandfathers of all, according to the Cree. I can imagine how dismissive early missionaries must have been of such "backward" notions. But it doesn't take many windbound days on the lake to begin to think like the Cree.

Leaving packs beneath an overturned canoe close to the water line may be acceptable on small lakes but is foolhardy on a lake like Winnipeg. As wind blows continuously from one direction over a lake, water can accumulate at the downwind end. In most lakes the rise is too small to notice. But a strong wind blowing across a very large open lake can create a wind tide of sorts that can raise water levels several feet. Ignorance of that risk can have consequences: soaked food packs.

Lake Winnipeg's waves are steep. Frances Russell writes of the folk wisdom of early Icelandic settlers on Lake Winnipeg that warned of the Three Sisters, Agnes, Mabel, and Becky, three very large waves that aggregated in stealth to rise suddenly and sink boats. Instinct argues hug the shore, but when waves are ferocious, breaking waves and bounce-back currents make near-shore the least safe place of all.

Geri and I wait windbound on Limestone Point. Hours pass. The sounds of Grandfather Wind and Grandfather Wave wander the corners of my mind. We should have taken the short portage marked by the caribou skull at the narrow base of Limestone Point north across to Limestone Bay, and let the point shield us from the big lake. But these breaking waves— wow, what a spectacular waterfront!

We explore the woods across the peninsula to the Limestone Bay side. Gentle waves lap at the shore, and we decide to cut a trail and carry our gear across. In an hour and a half we launch onto the bay, make five miles an hour down the shore, and camp on sand at the point's western tip.

Two days later, we start our final twelve miles paddling on glass. What a spectacular sight, the great lake as calm as a beaver pond. We savor the Gray Lady's parting gift.

·   ·   ·

How incongruous that such a large and powerful lake, that can scare the bejabbers out of boat pilots, canoeists, and fishermen alike, can only depend for its future well-being on decisions and actions taken not just on its shores but thousands of miles away.

Seeking Hard Bottom

The traveler asked the boy if the swamp before him had a hard bottom. The boy replied that it had. But presently the traveler's horse sank in up to the girths, and he observed to the boy, "I thought you said that this bog had a hard bottom." "So it has," answered the latter, "but you have not got halfway to it yet." ... So it is with the bogs and quicksands of society; but he is an old boy that knows it.

HENRY DAVID THOREAU

An eroded sandy path leads past red pine and scattered aspen down a steep slope to the lake. The trail ends at what one might expect to call the lake's shore. I choose not to use the word "shore" to avoid creating false impressions. No waves lap sand or stones here, nor bend the stems of rushes, nor ruffle lily pads. A fringe of alder and willow marks the edge where path meets spongy black mud. Open water lies half a football field away. A spongy, wet meadow lies between.

A trail of flattened sedge and an irregular string of stepping boards leads from the edge of the alder through hillocks of grasslike sedge to the edge of open water. I walk carefully, picking my way from board to board, toward the lake, Geri right behind. Soft squishy mud oozes up the edges of the boards with each step. Many boards lie beneath the oozy black, forcing me to guess where each hides. I take a misstep and my leg sinks to the shin. I understand why few people visit places like this. For them perceptions of such places will be constructed from fleeting views while passing by on roadways. Geri and I have never visited this lake. We have come to investigate its prospects as a place to gather wild rice.

The mud-boards end at a docklike assemblage of weathered slats and plywood held a few inches above open water by posts driven into the muck. A small aluminum fishing boat lies upside down in the sedge clumps to the side.

· · ·

Alongside the dock the lake bottom appears half a foot down, shallow enough that presumably one could wade into the water and load a canoe floating on its surface. I once made such a mistake on a similar lake. My partner and I had paddled to the edge of what looked like a shallow shore. I hopped out of the canoe into the apparent foot-deep water to pull the canoe onto land. I sank instantly to my hips and, descending rapidly into thick black soup, grabbed a sedge clump and wriggled my way out like a waterlogged muskrat. Putting a foot into this stuff is like stepping into a bottomless pile of black down.

Limnologists call this dark ooze "gyttja" (yit-ya). Canoeists and duck hunters, who know it as "loon crap" and by other more earthy names, may not realize how appropriate their scatological name is for the stuff. Limnologists describe gyttja as coprogenous, a fancy word for dung, excrement, feces. Gyttja is composed of the decayed remains of plant and animal bodies that have passed through the gut, not only of loons but of untold numbers of other creatures great and small. Bacteria and fungi convert such matter into this soft, black ooze, the keystone in the grand cycle of nutrients and flow of energy on which life in the lake depends.

Viewed from the dock, the open water occupies only a modest portion of the lake basin. Not far out, rushes and lily pads emerge out of the black soup. Beyond the lilies, clusters of tannish-green plants with long ribbon-like leaves and robust stems bearing heads reminiscent of grain coalesce in the distance into an unbroken vista of plants. *Zizania aquatica,* wild rice.

· · ·

The receding glacial ice left a poorly drained topography in its wake and millions of acres of shallow, sluggish lakes and ponds, weedy with rice and other plants. Such slowly moving waters are ideal conditions for creating the gyttja in which wild rice thrives.

Native wild rice once ranged from New England through the Carolinas to Florida, from the Dakotas through Iowa and Nebraska to Louisiana. Indians once found it abundant along the Potomac River and the Delaware River below Philadelphia. But it was in the glacier's backwaters of southern Manitoba, parts of Ontario and Michigan, and particularly Minnesota and Wisconsin that wild rice grew in large expansive stands to create the "wild rice bowl" of a continent.

· · ·

A thin stand of rice grows close to the dock, and I see dark elongate grains within the seed heads. As rice kernels mature they turn from milky green to dark brown. Many of the kernels have not yet matured. Rice on any given stalk ripens over a number of days. Should we harvest today, gather rice that is mature, then return every few days to complete the harvest? Or should we wait until nearly all the grains are ripe before picking? Mature rice falls

Wild rice, a highly nutritious grain, has disappeared from much of its former range.

easily into the water. Even a gentle breeze can send ripe grains to the lake bottom in an instant. To wait risks that wind blows the grains into the lake before we return. We decide to begin, and turn to renegotiate the stepping boards back to dry ground for our canoe and the ricing tools native peoples have used for a millennium: a long pole and a pair of tapered wooden flailing sticks. I take my place standing in the stern with sixteen-foot pole in hand. Geri settles onto a pad on the canoe's floor directly in front of me, a flailing stick in each hand, and we are under way.

As poler, my tasks are clear: spot areas where ripe rice is thickest; propel the canoe through the bed at a steady speed, fast enough to keep Geri's flails in constant motion knocking rice into the canoe, but not so fast as to cause her to miss some of the plants; then guide the canoe back and forth through the beds in parallel tracks six feet apart to maximize harvest. I spot a large dense expanse of rice to the north, shove the pole into the gyttja and push off. We are the only people on the lake.

In good beds, the best flailers, with strong wrists and quick hands, can harvest well over 400 pounds of rice in a day. For them an aggressive reach with one flail draws an armful of rice stalks over one side of the canoe. Then, two quick forward thrusts of the flail in the other hand, fast enough to make the air go *whoosh whoosh . . . whoosh-whoosh,* strip rice from stalks into the canoe. Flailers' sticks work alternately one side of the canoe then the other in rapid, fluid motion. We are not as practiced as the experts, nor as strong, and would be pleased with sixty pounds for the day.

As we approach a dense stand, a large flock of blackbirds rises as one out of the rice, flies low over the bed, then drops out of sight into rice farther away. Blackbirds and other song birds feed on rice. This aquatic grass deserves its alternative name of "blackbird oats." Unfortunately, from our point of view, birds landing on rice plants often dislodge many ripe grains, sending them to the lake bottom, retrievable only by ducks. The bed is a patchwork of dense and less dense plants. I determine a line of travel that will take us through the thickest stands and resume poling.

Geri begins swishing her flails in earnest. Rice falls into the canoe in a cadence of bursts, the delightful sound of intermittent rain, as she works one side then the other. Her rhythm, and so her success, depends entirely on my success propelling the canoe. The density of the plants renders paddles all but useless in effectively moving a canoe. Only a long pole is equal to the task. Although the lake here is no more than four feet deep, thrusting a pole into this bottom is akin to pushing one's way across a bottomless bowl of porridge. My rice pole has a duckbill at its end, a simple device with a pair of eight inch long hinged aluminum lips that spread widely apart when shoved into the mud, allowing me to get a semblance of purchase on the "bottom." The lips collapse back together when the pole is lifted out of the muck.

My poling is not producing the steady rate of speed Geri needs for efficient flailing. The canoe comes to a dead stop too often, and sometimes is even drawn backward, as I extract the pole from the mire to push off again. In fifteen minutes we come to the end of the bed. We must reverse direction. I plant the pole to the side and complete an acceptably tight pivot turn so as to make another pass through the rice parallel to the first. This maneuver, not carefully done, puts the standing poler as well as the canoe and its contents at risk of tipping over.

Years ago I poled for an expert ricer. Todd regularly harvested rice in quantities that Geri and I only dream of. Aware of my limited experience, Todd was blunt about my responsibilities: "If you ever sense we're about to capsize, jump out, we cannot spill the rice into the lake." Twenty minutes after we began, my misplaced step produced the telltale lurching of a canoe starting to roll. Like the dutiful soldier I leaped over the side, plunging through the water into the soft black ooze below. I popped back to the surface, pants and shoes, socks and underwear all oozing black gyttja. I crawled over the gunwale into the canoe, where Todd and the rice waited unscathed, and removed a strand of coontail dangling from my dripping glasses. "Nice job," he said. Ten minutes later it happened again.

Our turn completed, we enter a large patch of rice, several feet taller and more robust than we've seen. Even the grain heads are larger, bearing more kernels. I expect the rain of rice into the canoe will become a downpour. It doesn't happen. The stiff stalks are hard to bend and are so long the rice grains hang over the other side of the canoe. Poling becomes harder as well. We lurch forward in short jerks. Soon it takes all my strength pushing on the pole to merely inch forward. We finally clear the last of this tall dense rice and conclude that thinner rice will be more productive.

Distractions come easily while standing in a canoe in a rice bed. A small chicken-shaped bird flushes from behind a plant and flees ahead of the canoe, skipping nimbly from stem to stem then disappearing into the rice. Wetlands and shallow rice lakes, abundant in plants, provide excellent habitat for many birds: coots, moorhens, and rails among others. Rails in particular are secretive birds and, given their preference for these dense wet places, are seldom seen. Poling a canoe through rice beds significantly increases my chances of seeing them. I'm soon trying to convince myself I've just seen one, particularly since I'd heard the *keek* call of a rail earlier in the morning. Rails are exquisitely adapted to live in densely vegetated places. Their bodies are laterally compressed, not as much as in sunfish, perhaps, but compressed enough to allow them to slip silently through narrow passageways among the plants without a rustle to alert predators to their whereabouts. To see a frontal view of these birds is to instantly understand the origin of the phrase "skinny as a rail." They also can hold their feathers tight against their bodies to aid stealthy passage.

. . .

Standing, I can determine which rice beds appear most promising, but Geri senses more accurately how quickly or slowly rice is actually entering the canoe. "I'm getting almost nothing from these plants. Have we been through here before?" she asks. Distracted by the birds, I have let the canoe wander off course. I plant the pole sideways and slide us out of already harvested plants.

As I turn the canoe at the end of a run, the duckbill has ensnared and yanked a rice plant from anchorage. The plant now acts like a sea anchor, slowing the canoe until I jiggle the pole, loosening the plant, allowing it to float to the surface. Long air-filled chambers fill the stem, making the plants buoyant. This great buoyancy enables the weak stems to grow and remain upright in the water. It also makes these plants extremely vulnerable to sudden increases in water level. A six-inch rise in lake level in June or July, when the growing plants have just reached the surface, can easily uproot them and devastate a rice crop.

The rice thins as we complete our next run, and we decide to go ashore, stretch our legs, and have lunch. I redirect the canoe into a shallow lagoon. Ahead of us lies a mudflat and clumps of sedge marking the edge of drier ground. The canoe floats on three inches of water which lies atop fluffy black gyttja the consistency of tomato juice. As I pole toward the mudflat, the water layer thins and the gyttja thickens; black juice becomes watery black porridge and finally thick black pudding at the edge of the mudflat. The gyttja surface appears dry enough to walk on. We know better. I strain on the pole, plowing the canoe forward through the pudding, aiming at several sedge clumps that appear firm. A last power thrust and the canoe becomes stuck two feet short of the intended sedge dockage.

Geri clambers onto the bow plate and leaps toward the clump of sedge. Her leg sinks to the knee as she grabs hold of a tuft of sedge and pulls herself out of the mire. I toss our lunch sacks to shore then step onto a broken tree branch lying on the mud and totter to join her, muddying a tennis shoe.

We eat sandwiches and survey scattered rice plants growing robustly out of the mud in front of us. Few people, I expect, would judge this land-scape as appealing. Mudflats. Horizon obscured by clumps of coarse plants. Many might call it ugly. Yet despite our inconvenient landing spot, hindered views, muddied shoes, and the expected struggle to get the canoe back onto water, I feel good, even comfortable here.

We gather empty lunch sacks and water bottles, and I lean out, grip the bow plate of the canoe, pull it to us, and we get in.

Unfortunately, a wind developed as we ate. It now blows sideways against the canoe, sliding us downwind, perpendicular to our previous line of travel. The sliding canoe submerges rice on the downwind side before Geri can bend the stalks over the gunwale. On the upwind side, stalks slowly emerge out of the water after the canoe has passed and are unreach-able. I change direction to run the canoe directly up- and downwind.

We now cross our morning tracks rather than parallel them. The rain of kernels into the canoe stops abruptly each time we cross a previously traveled path. *Rain ... Silence ... Rain ... Silence ... Rain ... Silence.* At least we are not contending with blustery winds that strip the beds, snatching away hopes for any harvest at all.

We complete our final run of the day, and tired arms exchange pole and flails for paddles to cross the open water between us and the dock. A small fishing boat pulls up minutes after we do, and two older men step onto the wooden slats, fishing rods in hand. "Any luck?" I ask.

"Not bad," replies the older of the two, who looks to be in his eighties. His partner holds up a heavy stringer of crappies and perch and several nice northern pike.

"Wow," exclaims Geri. "I didn't think there'd be any fish in a lake this shallow."

"Oh, yes," responds the older man. "I've been fishing here since the thirties. There are fish here, all right."

We load rice into bags, gather our belongings, and trudge the mud trail to make room for the fishermen on the tiny dock. We drive off, heading to our primitive camp in the woods several miles away.

· · ·

Near camp, on a rutted forest road, we encounter an Indian man in a red pickup coming out. We exchange waves. I expect he is headed to town for supplies. His family preceded us to the informal campsite on a pine-covered upland overlooking an extensive wetland. A small river wanders a sinuous path, looping back on itself as though befuddled about where to flow. Our tent sits in a grassy clearing that extends to the edge of the marsh. A fringe of rice bounds the edge of the sluggish water as it slips out of sight behind cattails. The bank rises up from the landing forming a small hill.

A large red pine, much older than the other trees, grows on the slope facing the marsh close by the Indians' camp. Seen from here, the wetland widens considerably in the distance, creating several small lakes that appear as slivers of distant reflection. Hues of green and ochre, counterpoint shadows, glistening spots of reflection, and horizon lines converge in the distance.

· · ·

I expect our Indian neighbors perceive this landscape differently than we do. As the only native cereal grain to occur in great natural abundance in North America, wild rice has long been an important food for native people. Richer in protein, vitamins and minerals, and an essential fatty acid than many commonly farmed grains, wild rice is excellent food.

Wild rice, "manoomin" to Ojibwa people, assumed much greater significance in Indian lives than simply food. Eva Lips spent time with the Ojibwa at Minnesota's Nett Lake, among the finest rice beds anywhere, and concluded that for the Ojibwa people the harvest of wild rice "was the decisive event of the year, of the total economic life and with it, life itself." It permeates the legends, stories, religious and medicine ceremonies, and spiritual beliefs, attaining near sacred status. Other-than-human persons held in their hands the fate of the rice crop and the safety of ricers. They could, without propitiation, destroy a crop or cause ricers to drown. Rice was buried with the deceased, or placed beside graves, as nourishment for the spirit's journey into the afterlife. Indians also periodically returned to gravesites years after a burial to leave additional rice.

Conflict over ownership of the rice stands and the shallow lakes that contained them created near constant warfare between Dakota and Ojibwa peoples. Rice lakes became battlegrounds. Later, when reservation boundaries were imposed, Indians pleaded for rice lakes to be included within them. While those wishes were often accommodated, in cases where beds were initially excluded, Indians protested vehemently, offering to give up acreage elsewhere on the reservations in trade, if they must, to secure the all-important rice lakes.

. . .

I empty our rice onto two tarps, spreading the grains uniformly a few inches deep. A freshly spread surface of newly harvested rice usually comes alive in riotous motion produced by a miscellany of tiny crawly creatures, among them rice worms, immature moths that feed voraciously on rice flowers and, in mature stages, eat the rice kernels themselves. Aphids and larval insects that feed on tissues inside the stem are usually present as well. Our rice is nearly motionless; only a few tiny worms and some dainty cream-colored spiders wander its otherwise peaceful surface. Subsequent processing to remove chaff from the kernels also removes the animal life.

I look out again over the marsh and scattered patches of rice. Many eyes would see this wet landscape as forbidding, as it offers no solid ground for humans to comfortably cross. Much of the vegetation is too thick for a canoe, except for the narrow ribbon of stream, and even the stream is mostly confined within dense green walls of vegetation. Yet, as at our mudflat lunch spot, feelings of pleasure and contentment return. A perception conflict arises in my brain. My eyes tell me unattractive views surround me.

However, if perception of place is based purely on visual image, why am I so comfortable here? Landscape image is not landscape *truth*. My mind must be merging the mud with the physical pleasure of poling, our intimacy with the rice, the animals, and the water, and the delectable sounds of brown rain accumulating at our feet. My mind seems to be replacing the sense of *landscape observed* with the different truth of *landscape experienced*. I think how, except for shores and surfaces, we are unable to experientially perceive, and therefore understand, the greatest portion of most lakes.

What must be the experience of our Indian neighbors as they look over this same space from their encampment? How many generations of their people has the old red pine watched return here during the moon of the rice harvest? How many of their ancestors has it watched fight, and even die, for rice beds on this very ground? That there was such fighting here I have no doubt. A now discarded early name for the lake we harvested today was "Battle." At home I have a black stone of my grandfather's that may have participated in some of these fights. It is fat, ellipsoidal, three inches long, and two and a quarter inches wide, a very dense stone. Its surface is smooth except for a roughened depression the width of a little finger that runs around its middle, by which a handle can be attached by thongs to make a war club. Grandpa was a softhearted immigrant from Norway who lived his adult life not far from here. How Ole came to own this weapon of war I cannot guess. But what a find it was for the warrior that first pulled it from an ancient shore and chipped out the groove to attach a handle, creating a powerful weapon for defense of the rice lakes.

Rice crops are inconsistent. Bumper crops appear only about one year in four, the same frequency as crop failures. Failed crops were calamitous for Indian peoples, and early accounts describe great suffering, even starvation, in the bad years. How many times over the centuries has this pine tree watched Ojibwa people dismantle camp and head to wintering grounds carrying the paltry harvest of a failed rice crop? How often has the tree seen the future in pensive eyes and drawn lines in the weathered faces of people who understood all too clearly that their skin sacks held nowhere near enough manoomin to see families through until spring.

Since individual perceptions are highly personalized creations, constructs unique within each person, I cannot hope to understand what our neighbors see when they look out from this red pine, filtered as their view must be, through the lens of centuries of tribal memory, centuries of lives intimately interwoven with this landscape. I can only speculate at the depth, richness, and meaning of this landscape to them. How different their feelings and mine must be from those who know this landscape only as scenery in passing, whose minds have only visual images from which to construct a perception.

The shallow rice lakes were important to early Europeans as well. Fur-trading posts relied heavily on rice for winter provender, obtaining it by barter with the native people. Explorers, fur traders, and early settlers all carried rice with them into the wilds. Had rice as a wild crop been reliable in quantity from year to year, and had rice grains the habit of ripening all at the same time, enabling efficient harvest, agricultural history and the fate of many rice lakes and associated wetlands might have played out differently than it did.

Instead, the growing wave of settlers experienced these shallow lakes and wetlands through eyes that saw such places as worse than worthless, as impediments to agriculture, and without redeeming value.

St. John de Crevecoure, in his book, *Letters from an American Farmer,* published in 1782, anticipated the future of wetland history in North America when he informed his readers that the next thing a farmer carving out a place in the wilds does, after chopping down all the trees his muscles can manage for the year, is turn to improving his land by draining away its wet places. The shallow lakes and deep marshes that created rice habitat were among the millions of wetland acres spread widely across the country at the time of de Crevecoure's book. Over half of that acreage has now been drained, and wild rice has disappeared from many of its former locales.

The smell of smoke from our campfire where Geri is cooking supper interrupts my thoughts. I am famished and follow the tire track down the slope to a steaming bowl of soup and, soon after, nestle weary muscles into a soft sleeping bag in the tent with my wife.

. . .

We awaken to thick morning fog on this our last day of harvesting. We drive to our lake, and the fog lifts suddenly as we arrive, bathing the lake in blazing morning sun. The air is warm and calm as we launch and head back to the north beds. From a distance the heads appear much darker, riper, than before. Even thin stands today yield bursts of brown rain. We spend most of the morning here. At times the picking is very good. I finally see two rails, skinny as can be, and now swans and a muskrat. Muskrats are nearly as fond of wild rice as humans and can consume a sizable part of a crop.

We eat lunch in the canoe to save time. Rice covers the canoe bottom half a foot deep. Newly falling grains plow head first into the pile with their needlelike awns sticking straight up, creating a furry surface. What deep joy to move a hand through the piled rice and feel the heft of the grains. Dozens of tiny, translucent spiders busily construct webs, seemingly indiscriminately, from my knee to the pole, a lunch sack to my shoe, from Geri's elbow to the gunwale and even to a rice plant outside the canoe.

In early afternoon we face the age-old dilemma of ricers and berry pickers: when should one leave one's current spot with its reasonable picking to check a new place where picking might be even better—or much worse, the exploration an abject waste of time and foregone rice? With an hour left in our season we strike out for the east-central beds and find the ricing there the best yet. Why did we wait so long to move?

"It's three o'clock," Geri announces. Legally, we must stop. We lean back and breathe deeply. Ricing is hard work and muscles are tired, but in a highly satisfying way. "What a fantastic day. What a gift to be here," Geri continues, mirroring my own deep joy. For some, like our friend Todd, the ricing experience is deeply spiritual. "This is where God lives," he says. Enveloped by my own feelings of peace and connectedness to the cosmos, I agree.

I reluctantly stow the pole in the canoe and we paddle slowly across open water toward the dock. We land and begin carrying our belongings to the car. I hesitate on the weathered dock and look back at the lake. Most people, other than the likes of duck hunters, ricers, and birders, will never experience these weedy lakes and deep water marshes as we have. For them, places like this can never be more than visual image. But if landscape as experienced feels more trustworthy, more valid, which landscape experience, whose landscape experience, is truth? Or is that question impossible to answer, a query of fantasy, swamped in a lakebed of the amalgamated gyttja of individualized experiences? Where is the hard bottom of truth? The answer matters. Landscape perception shapes values and determines choices and behaviors.

Other truths are more accessible. The boy in Thoreau's story of the mired horse and rider is surely correct; there is a hard bottom of established truth about the depth of the mire. A pipe driven into the soft gyttja of a rice lake in Wisconsin found hard bottom twenty-three feet below the water surface. The first three of those feet were lake water and the remaining twenty feet were gyttja "bottom." Twenty feet of a lake's life used up, three feet left to live. But a pipe and driving maul are no help in untangling the complexities of landscape meaning. I pick up the pole and our bag of rice and step onto the trail to the car. We leave the lake to break camp, load our ninety-six pounds in the trunk, bring our rice to a processor to remove the chaff from the grains, and head for home.

. . .

Our part of town was once sprinkled with Indian mounds. Most of these burial places were destroyed before such sites became protected by state and federal law. Two of the few that survive lie a hundred feet from where I write these words, in a strip of woods behind my house. A block away, Elm Creek carries the waters of Hayden Lake to the Mississippi a mile down-stream. Though smaller than the lake we riced, Hayden Lake is otherwise much the same: shallow, bordered by a fringe of plants rooted in rich black gyttja. Though protected as parkland, no paths reach the water's edge. The lake is nearly inaccessible, forgotten by humans, though not by birds and a beaver family that lives near the lake's outlet. A few years ago, in early spring before ice-out, I explored the lake's perimeter and, in one isolated spot at the west end of the lake, discovered a few spindly rice stalks protruding above the ice, the paltry remnants of stalks too short to harvest from a canoe. I expect these plants are lingering descendants of a grand lineage once abundant here.

I inquire of the park naturalist what she knew of wild rice in Hayden Lake. "I've heard that years ago there was rice there, but I really don't know anything about it now," she says. "Basically, nobody goes there anymore." I wonder how, in life, the souls long buried in the mounds behind my house would have responded to those words. I feel a strange compulsion to carry a handful of our rice across my backyard into the trees and leave it for the Indians buried there.

MINDSCAPES

Blue-Green Nemesis

Everything has indirect effects and interconnected functions, so if you alter one process, you are bound to disturb a thousand other processes you did not even know existed.

NANCY LANGSTON

Today, August 28, marks the one-year anniversary of my epiphany visit to Diamond Lake. Curiosity has brought me back. Has the lake's condition changed since my mind-shaking visit? From the landing the lake appears much as it did then, a sea of blue-green algae run riot, as dominating and repulsive as before. As then, lake bottom disappears from view in shin-deep water.

Detestable though blue-greens are in such crushing numbers, I admit to grudging respect and intense curiosity about any living being that can so overwhelm a space. Who are these creatures and what enables them to become so dominant?

Before I launch the canoe to learn what may have happened elsewhere in "their" lake and its bays in my absence, I must tell a story. I must tell a story about what I have learned about these blue-greens. Together with their sidekick accomplice, the paradoxical element phosphorus, they make a remarkable and fascinating tale. Hearing this know-the-enemy story will help explain just what Diamond Lake and other pea-soupish lakes are up against.

Mostly too small to be seen by the naked eye, blue-green algae (cyanobacteria) can overrun a lake rich in phosphorus.

. . .

My students and I have examined blue-greens under a microscope. We have seen them as colonies of round or box-shaped cells, some in small clusters encased in transparent sheaths, others as bluish necklaces, twisted spirals, even straight filaments of cells lined up end to end like a row of bricks. I have seen some form rubbery nodular colonies as large as golf balls.

Blue-greens, once considered a kind of algae, have been discovered not to be algae at all, but a more primitive form of life with an impressive history of transforming things here on earth. Technically, they are known as cyanobacteria (*cyano* means blue in Greek), though force of habit leads many people to continue to call them blue-green algae. By whatever name, they are the oldest life forms known. Long before the Age of Dinosaurs there was the Age of Cyanobacteria, when for 2 billion years these simple organisms alone inhabited our planet. The earliest ones apparently released much oxygen into our atmosphere, making earth inhabitable by creatures like us. Later in earth's history the blue-greens helped transform dead plants into earth's crude oil. They thrive today in lakes, in hot springs, hot deserts, polar deserts, glaciers, and even the saline lakes and ice sheets of Antarctica. In places they multiply to reach the incredible density of 400,000 cells in a single drop of water. I grudgingly tip my hat to them.

Reproductive success boils down to obtaining enough food and avoiding being eaten long enough to have babies. Blue-greens are devilishly good at both. Waterfleas and other microscopic animals comb lakes in search of food the size of algae and cyanobacteria. Many cyanobacteria, unlike most algae, produce toxic chemicals to ward off such predators.

Perversely, cyanobacteria toxins are more poisonous to wildlife, farm animals, and humans than to aquatic life. Deaths result when animals drink in cyanobacteria in great numbers or eat cyanobacterial scum from the water surface. Animal deaths caused by cyanotoxins are reported from many states and provinces, with cattle, pigs, horses, dogs, chickens, and turkeys appearing most frequently in these reports. But cats, squirrels, fox, hawks, skunks, mink, snakes, frogs, salamanders, wild geese and ducks, gulls, songbirds, even honeybees and, in South Africa, three rhinoceroses have succumbed to the toxins of the giant-slayer cyanobacteria. A local veterinarian described the death of a dog from blue-green algae poisoning here on Diamond Lake as gruesome.

This type of cyanobacterium can produce deadly toxins that are able to kill many kinds of animals and sicken people.

Experts long believed few cyanobacteria species produced toxins, did so infrequently, and the overall risk was minimal. That changed years ago when Brazilian hospital patients became ill. Some people ultimately died, and all the illnesses and deaths were traceable to cyanobacterial toxins from a contaminated water supply. Illness due to liver damage is a more common risk to people than death. Thankfully, not all cyanobacteria blooms on lakes are toxic.

In one small triumph against these miniscule giant-killers, several fish species around the world have developed immunity to the toxins and gobble up cyanobacteria like gourmands feasting on New England clam chowder.

Paradoxically, some cyanobacteria species have a long history as human food and have been sold as a health food in America. Health officials worry because some of the cyanobacteria used for this purpose are obtained from natural blooms in lakes and, unless subjected to complex tests to verify their safety, could present considerable risk to those who eat it. I am puzzled. Cyanobacteria are noted for causing repulsive tastes and odors in water supplies that typically put people off. Either processing destroys the sensory warnings or natural food aficionados judge that if it tastes bad it must be good.

Despite their nasty poisons, however, blue-greens could never achieve the astronomical numbers that live in this lake without the help of phosphorus. Of the dozens of kinds of atoms that compose life forms, surely phosphorus is among the most enigmatic. The superstitious would say, what else would you expect from the thirteenth chemical element to be discovered? The squeamish might say, what else would you expect from a substance discovered by an alchemist seeking to make gold from putrefied urine?

Consider its incongruities. Phosphorus spontaneously ignites in its elemental form, explodes in air, and must be stored under water. It is also a constituent of some of the most flame-retardant substances known.

Phosphorus is one of a handful of elements that are critical ingredients of the most important molecules of life. Without phosphorus there is no DNA blueprint to guide the development of life forms from bacteria to fishermen. Yet it is also a critical ingredient of some of the most hideous agents of war, deadly nerve gas and phosphorus bombs, the wretched incendiaries that roasted humans alive in World War II.

*Phospho*lipid molecules form the molecular "skin" of cells. No phosphorus, no cells. No cells, no life. Yet imagine. This element is also in some of the most deadly pesticide poisons known. Without phosphorus, lakes are as lifeless as the moon, yet excess phosphorus can ravage them.

· · ·

The early alchemists, of course, knew nothing of phospholipids or DNA as they hunched over workbenches cooking pots of aged urine, oblivious to what actually lay at the bottom of their boiling kettles. Boiling the golden water in just the right way, they believed, would produce gold.

We chuckle at their naïveté, at the image of such people, faces lit by a flickering fire, toiling at their benches amidst flasks and bottles of aging urine, foul odors wafting about, all in pursuit of the riches supposedly so close at hand. But, is the idea of transforming insensate atoms into the flesh of a perch or a prince any less cockamamie? Bind atoms of phosphorus, oxygen, hydrogen, carbon, and aluminum in the proper ratio, and you end up with a crystal of turquoise, dumb as stone. Bind the same atoms of phosphorus with carbon, hydrogen, oxygen, nitrogen, and a few more, and you have crayfish and poets. Life, even a single cell of the simplest bacterium, invisible to the eye, towers in complexity far exceeding in eminence a Mount Everest in its eternity of stone. Life: alchemy at its zenith, with carbon, hydrogen, nitrogen, oxygen, and phosphorus at its core. The alchemists never could have imagined that the phosphorus they found in the pots of cooked-away urine was vastly more potent than the gold they sought, a piece of the philosopher's stone.

I cannot directly "see" phosphorus in the lake here. The world of atoms is far too small for human eyes to discern, but its pea green consequences stretch from shore to shore.

· · ·

Cyanobacteria cannot conjure up their obnoxious blooms out of nothing. Like people and perch, they must eat. All obtain phosphorus and other elements we need from our food. The diet of cyanobacteria, and all photosynthetic life, is far simpler than ours. Plants, algae, and cyanobacteria spin the simple materials of water, carbon dioxide, phosphate, and such into the golden carbohydrates, proteins, and fats that compose themselves and feed the rest of us.

Creating life's stuff is much like another bit of alchemy I'm fond of: making blueberry muffins. Surprising, perhaps, but the parallels are real. This delicious metaphor illustrates what I mean. Geri's recipe for those yummy golden brown muffins, spotted with luscious marbles of blue, calls for two cups flour, ¾ teaspoon salt, ¼ cup sugar, two eggs, two tablespoons butter, ¾ cup milk, one cup berries, and two teaspoons of baking soda. This recipe yields two dozen delectable muffins. I can scarf down a dozen of them at a single sitting, to the dismay of other members of my family.

The solution, of course, is to multiply the recipe and make dozens and dozens of them so all can have their fill. Out come the mixing bowls and canisters. Ingredients are measured and mixed. But the baking soda tin is nearly empty, only two teaspoons worth to be had. We can only make one recipe. The pantry may bulge with barrels of flour and sugar and salt, the refrigerator overflow with tubs of butter and gallons of milk, crates of eggs, and berries. All superfluous. The paltry two teaspoons of baking powder stop the show.

Algaeplasm has a recipe too: 106 parts carbon, 16 parts nitrogen, 8 parts calcium, some potassium and magnesium, sodium and sulfur, and miniscule amounts of a dozen more, and one part phosphorus.

Algae, aquatic plants, and blue-greens create the golden molecules of life at a prodigious rate, as long as the ingredients hold out—which they do not. And in lakes, except in unique circumstances, the "tin" of phosphorus usually empties first. Compared to demand, it is phosphorus that is available in least supply, the bottleneck to alchemy. Little phosphorus in lake water begets few cyanobacteria, algae, and aquatic plants. Lots of phosphorus begets lots of blue-greens or aquatic plants or both. The cyanobacteria explosion in Diamond and other pea green lakes reflects a richness of phosphorus beyond normal limits that tips the competitive balance to favor cyanobacteria.

· · ·

It would have surprised Henry Thoreau to learn that the substance used to make the newfangled matches of his day, this phosphorus, was one and the same as that accounting for the substantial differences he observed among the lakes around Concord. His comparisons of Walden Pond and Flint's Pond readily express the differences between lakes where phosphorus is plentiful and lakes where it is not.

But a digression is in order. To Midwesterners a pond is little more than a glorified puddle, shallow and choked with plant life. Half-acre storm-water catchment basins are ponds. You will understand, then, my surprise on my first visit to Walden Pond. I did not expect to find a sixty-acre *lake*. And nearby Flint's Pond is over three times as large. What actually is the difference between a pond and a lake? Limnologists think of a lake as a place wind plays a major roll in mixing the water column. Ponds are water bodies where temperature more softly drives the mixing. Regardless, I expect New Englanders will stand by their word "pond."

Hear the distinction Thoreau makes between Walden Pond and Flint's Pond.

> [Walden] is a clear and deep green well. . . . The water is so transparent that the bottom can easily be discerned at the depth of twenty-five or thirty feet. Paddling over it, you may see many feet beneath the surface schools of perch and shiners, perhaps only an inch long. . . . It is nowhere muddy, and a casual observer would say that there were no weeds at all in it. . . . A closer scrutiny does not detect a flag nor a bulrush, nor even a lily, yellow or white, but only a few small heart leaves and potamogetons. . . .
>
> Flint's, or Sandy Pond, in Lincoln, our greatest lake and inland sea, lies about a mile east of Walden. It is much larger . . . and is more fertile in fish; but it is comparatively shallow, and not remarkably pure.

Walden's greater clarity in contrast to Flint's "not remarkably pure" waters in Thoreau's day resulted because it had fewer algae and cyanobacteria to obstruct the penetration of light. Fewer algae mean less food entering the food chain. No wonder Thoreau found Flint's Pond more fertile in fish. Just as that mostly empty tin of baking soda in our kitchen limited the number of muffins Geri could make, limited phosphorus capped the production of life in Walden. Phosphorus spelled the difference.

Availability of phosphorus to aquatic life is complicated by its vexatious behavior once it gets into the lake. One could forgive plants, algae, and cyanobacteria for concluding that the gods of phosphorus chemistry conspire against them. For starters, phosphates are drawn to particles of clay and certain other substances as iron filings are drawn to a magnet. These particles then sink out of reach of algae and blue-greens to the lake bottom below. Much phosphorus is tied up in the bodies of waterfleas and insects, plants and fish, in theory available for recycling upon the deaths of the individuals. But here too the lifeless forms and their phosphorus sink and merge into the bottom sediments.

Competition for phosphorus in a lake among individual algae, water plants, and bacteria is fierce, and adaptations to obtain it are finely honed. Cyanobacteria are sinfully good competitors. Their first strategy is disarmingly simple: keep the body small, very small. A bit of geometry explains. A mop composed of a hundred strands of cloth absorbs more spilled water than a mop with a single solid cylinder of cloth, because the many-stranded mop exposes more surface area contact for slurping than does the solid mop. The same principle is at work for small life. A hundred individual cells expose more surface to take up phosphorus than would be the case if the cells amalgamated into a single mass. The smaller the cell, proportionately, the greater the amount of cell surface available to absorb food compared to the body mass that must be fed.

We humans are instinctively familiar with another strategy: take in as much as you can, more than you need, while the getting is good. Here today, food or phosphorus may be gone tomorrow. Such luxury uptake and storage helps creatures survive times of lean with leftovers stored up from times of feast. During the fat times cyanobacteria take up enough phosphorus to carry them through several phosphorus-hungry baby-making cycles when no external phosphate is to be had. Comparable excess intake of calories produces unmitigated obesity in humans.

Some blue-greens set up housekeeping in the phosphorus-rich lake bottom sediments. Times for them are nearly always fat, at least for phosphorus. But light, another crucial ingredient in making vegetableplasm, is much diminished with depth. Unlike their algal competitors, many cyanobacteria have internal adjustable flotation bags to have it both ways. They add gas to the bags to make themselves lighter than water to float toward the surface for light during the day and then collapse the bags, making themselves denser, to sink to the sediments to take up phosphorus and other goodies at night.

. . .

Diamond Lake has succumbed to an unholy synergy. How did this lake come to possess such a mother lode of phosphorus? The pastoral view from the landing provides few direct clues. No factories belch smoke or pipe phosphorus-laden chemicals into the lake anywhere on its shore. I launch my canoe onto the blue-green rich water to discover what changes have occurred in the relationships among humans, lakes, phosphorus, and blue-greens since my epiphany visit. A single motorboat plies the middle of the lake; otherwise, I am alone.

Fifty yards west of the landing I notice the first change. No sickly brown-green plant mat floats where a particularly odiferous one lay in my last visit, though the density of blue-green algae remains high as before.

Farther down shore I realize I have not yet heard the pucker-smack sounds of carp, so familiar before. And no wonder. The green-brown stew of elodea and coontail choking the stems of the cattails that attracted so many fish before is no longer here. I see not a single elodea and only one tiny sprig of coontail. The bases of the cattails are naked. I can see through them to the damp sand on the shore. Waves now more easily attack the bank, washing soil particles with attached phosphorus into the lake.

Plant mat covers only half the area of the small south bay it did before. Even the duckweed is missing. I paddle easily across water I struggled mightily to penetrate before. I follow a great egret around cattails toward Cormorant Point. Someone has cut up and hauled away most of the large dead cottonwood that formerly attracted pelicans and cormorants. The old tree was a favorite rest site for the big birds. I see none today.

Plant mats in the next two bays occupy less than half the space they once did. Blue-greens are as dense as ever. In dying back, the massive plant mats have surrendered their phosphorus to the blue-greens.

A soybean field slopes to the lake, rows stopping two lengths of a paddle from cattails that mark the water's edge, as close as a tractor can get to the lake without tumbling in. On my last visit the crop was corn. What fertilizer the plants don't pluck up, runoff water will deliver unhindered into the lake.

A turtle sticks the tip of its snout above the water for a periscope look then submerges. Two coots explode out of cattails, feet paddling the water madly, then comically paddling air long after they become airborne. A flock of pelicans and another of gulls segregate themselves to share a narrow beach off a cattail island. I modify my line of travel so as not to goad them into energy-wasting flight. The pelicans eye me warily but hold their ground. The skittish gulls take wing en masse, circle twice, then return to their end of the beach.

Two grand houses come closer into view. The earth-tone structures sit in commanding positions upslope from the lake, each set amidst a vast manicured lawn. What wonderful views these owners have of the lake. The grass is the gorgeous green of plants well fed. A symposium speaker I once heard posed the question, "Which nonnative plant has had the greatest negative impact on our nation's lakes?" Images of the obnoxious invasive Eurasian watermilfoil and curly-leaf pondweed came to mind for most of the lake-savvy audience. With its shallow roots and thirst for fertilizer we homeowners readily provide, however, the answer is lawn grass. Thoughtfully, these homeowners have left a thick cattail fringe along nearly all their lake frontage that will retain some of the phosphorus and herbicides that runoff would otherwise carry hell-bent into the lake.

I paddle past the second grand estate. A wire fence runs into the water, separating lawn from a cow pasture. The shore here becomes steep bank. On previous visits black angus cattle had the run of this bank all the way into the lake. Much of the bank was bare dirt because plants are unable to cope with cattle hooves and erosion. Now, no cows huddle under the few shade trees still alive, no cows drop nutrient-rich cow pies on the sand or gush urine into the lake.

Today, a fence runs along the crest of the bank, keeping the black beasts up on the pasture and off the shore. Hallelujah! Whoops—a strong whiff of manure from the downsloping pasture wafts out over the water. But at least the shoreline is stabilizing thanks to weed growth and cottonwood saplings.

No sign of floating mats in the north bay, so I cut across its mouth to the row of houses along the north shore. Well-fertilized lawns run nearly to the water's edge. Years ago I ran for public office in the district that included that row of homes. As I went door to door soliciting votes, I listened to people's concerns. The most common complaint? The degraded condition of their lake.

"Diamond Lake had much clearer water when I first moved here," said some.

"It's that farmer's feedlot on the south shore that's ruining it," said others.

"Will you get the pollution control people to shut that feedlot down if we elect you?" asked another.

I looked at the luxurious, well-fertilized green carpet lawns that ran from most houses to the water's edge. Small capped pipes protruded a few inches above the grass in the yards marking septic systems buried below. I demurred, doubting the residents wanted to be informed of their complicity in their lake's problems. I am certain that lake property owners would not knowingly act in ways that degrade their lake. But urine flushed down toilets is urine out of sight, so out of mind.

Today no small white pipes protrude through the grass. Recently, at considerable expense to themselves, the enclave of homeowners on the north shore got their houses hooked up to sewer lines that now carry the waste to treatment. Most lawns, however, still run to the edge of the lake. Minnesota has banned use of fertilizer that contains phosphorus on lawns to reduce damage to lakes.

I paddle around a cattail point and into a nook of the northeast bay, always my place for final observations, thoughts, and field journal entries. I rest my paddle and sniff. Gone is the stench so pervasive in the epiphany visit. Gone also is much of the mat that once smothered this bay in coontail and elodea. Ten small clusters of lily plants decorate this part of the bay, all a robust green.

Families of geese float in the center of the main bay where I once came upon a hollow-eyed, shriveled pelican corpse sprawled on top of the mat. The geese notice me, begin honking calls of distress, and start flowing toward open lake. How do I tell them I mean them no harm?

I'm not sure what to make of the changes I'm seeing. Can I say the lake is improved? In ways it is. Travel is surely easier with much of the plant mats gone, but the massive blue-greens and their turbid dirty water counter the good. The lake remains highly distressed, but I've seen today the stirring of change. I leave the lake with hope.

. . .

Thoreau wrote, "You cannot see anything until you are clear of it." His admonition is spot on here. To fully perceive we must step back. Diamond Lake is not simply a sea of blue-greens washed in phosphorus. It is a sea washed in relationships with us.

Limnos III—Lake Mendota, Wisconsin

*Why do lakes differ so widely in productivity or in ability
to support a population of plankton?*

EDWARD A. BIRGE, 1911

Usually by November I have put my canoe away for the season, fearful of cold water that could quickly sap life away in a capsize. Today is an exception. Geri and I find ourselves slowly paddling under bright November skies on a lake famous the world over. Ostensibly, we have come to Madison, Wisconsin, to attend the annual conference of the North American Lake Management Society. The conference may have been merely an excuse to finally set eyes and canoe on Lake Mendota, birthplace of the scientific study of lakes in North America.

I swish my hand through water, surprisingly clear for being situated amidst a city the size of Madison. I have never been on this lake before—in body. But in truth it feels like an old friend. We slowly pass a string of elderly houses fronting the lake. How many times in years long gone by, I wonder, did the owners of these grand homes look out on the lake at a man in a boat dragging strange contraptions through the water? The man, with maybe a partner, would have been Edward A. Birge.

North American limnology was born here, and Edward Birge, assisted by Chancey Juday, was its father. Birge and Juday. Birge and Juday. As a graduate student at the University of Minnesota it seemed to me these men had written half the scientific papers I read. Incredibly, nearly a hundred years after their pioneering studies, I find nineteen citations of papers authored by these men, jointly or alone, in a leading college limnology text. Much of that research was conducted on this lake. As our canoe works its way west, the buildings of the University of Wisconsin, Birge's home base, become more distinct on the far shore.

. . .

After attending Williams College, Birge chose to pursue further studies under the great Louis Agassiz at Harvard. Only months after Birge arrived at the school, Agassiz died suddenly of a stroke. Birge continued at Harvard, where he became charmed, as have so many others, with the tiny inhabitants of a small shallow pond, those master seducers—waterfleas. His PhD thesis on daphnians became the first intensive study of this group. He left Harvard in 1876 for a teaching position at the University of Wisconsin.

. . .

We paddle past a park and watch geese walk single file between joggers and the shore. Ahead, numerous tall buildings, we assume dormitories or student apartments, rise not far back from the lake. Lake water turns less clear as we approach campus. Small clumps of cyanobacteria now appear.

Birge set out to investigate the distribution and behavior of plankton in Lake Mendota, and in 1895 published his findings on their vertical distribution in the lake. Several years later he hired Juday as his assistant. In six years the two men strained over half a million gallons of Lake Mendota's water to better understand the ecology and behavior of the plankton.

Birge's groundbreaking plankton work raised more questions than it answered. The two men, trained in zoology, soon discovered they could not come to understand lake life by studying lake life alone, and they became drawn into a wider lake world involving chemistry and the physical aspects of lakes. They learned that the amounts of dissolved oxygen in lake water changed with the seasons and varied with depth. They also recorded changes in water temperature from surface to bottom and coined the word "thermocline" to denote the stratum of lake water where temperature drops sharply from the warmer surface zone to the cold layer of the depths.

In their long careers Birge and Juday added immensely to our understanding of lakes. Birge conducted the first studies on the depth of light penetration and its consequences. They discovered that the amount of oxygen present in lake bottoms depended on the amount of dead plant and animal matter available to decompose there. Their full list of research subjects reads like a table of contents of a limnology book: organic content of lake water, heat budgets of lakes, wind's influence in thermal stratification, electrical conductance of lake water, concentrations of phosphorus and other minerals in lakes, and light absorption by lake water.

They discovered that photosynthesis, cellular respiration, and bacterial action worked together to create the distribution patterns of temperature and dissolved gases in lakes. Most important, they recognized their biological backgrounds were insufficient to enable them to adequately understand lakes, and they initiated collaborations with geologists and chemists and physicists. Through their work a lake became understood as not merely an assemblage of life forms interacting with each other in the sense of Stephen Forbes. Rather, a lake emerged as a biological-geological-chemical-physical system. They filled in the spaces of descriptive limnology, establishing a fundamental framework of how lakes work. Birge's biographer writes that the man's contributions to the University of Wisconsin ended seventy-four years after it began, fifteen months short of his hundredth birthday. Exploring why lakes differ widely in productivity or in ability to support a population of plankton became a lifelong adventure into unexpected realms for Birge.

. . .

New voices arose in lake science, particularly that of another lake studies giant, G. Evelyn Hutchinson at Yale. Hutchinson's extensive studies in biogeochemical systems greatly deepened the understandings developed by Birge and Juday. I have visited Linsely Pond, perhaps the best known of the lakes studied by Hutchinson and his students, drawn by its historical importance in adding to our perceptual understandings. The tiny body of quiet water gives no hint of its past significance, though when I was there several floats anchored at midlake appeared to mark sampling sites.

Among the last published reports of Birge and Juday, their study of a lake's energy relationships foreshadowed the next perceptual leap in understanding that was to result from a synergy between Hutchinson and a postdoctoral student from the University of Minnesota named Ray Lindeman.

· · ·

Now at the UW campus waterfront, we paddle slowly past university buildings. Flecks and gobs of cyanobacteria become more noticeable than before. We approach the last building fronting the shore before woods replace buildings—the Center for Limnology. The building has a garage-type door at the water's edge that, when opened, appears to allow a boat to actually enter the building.

I stop paddling. This is a personal moment for me. I gaze transfixed at the tan building and ponder the might-have-beens. I applied to graduate programs at two universities, Minnesota and Wisconsin, and was accepted at both. Though my undergraduate professors at Minnesota urged me to go to Madison to experience fresh ideas, a romantic interest that was not to pan out and uncertainty about the draft and Vietnam swayed me to remain at Minnesota.

What an unmatched setting here in which to study lakes. I didn't visit the campus before making my decision. Had I done so, how could I have turned away?

Who knows where paths not taken might have led. I do know that had I come here I never would have met Geri. Serendipity works in wonderful ways.

Diamond's Dot

What if all the ponds were shallow! Would it not react on the minds of men?

HENRY DAVID THOREAU

A map can show you where you are. Sometimes it stimulates questions about how you got there. I recently came into possession of such a map, one with multicolored dots scattered like a handful of spangles as an overlay on a map of the Twin Cities Metropolitan Area. Each dot represents a lake.

The map is a report card on the water quality of lakes tested the previous year. Dark-blue dots signify A grades, light blue ones Bs. Most numerous on the map are gray-dot Cs. Ochre dots are Ds; and, like school teachers' marks, bright-red F dots denote lakes that failed.

At the north end of my county lie two dots, red as sin. The upper dot is labeled Diamond Lake. I am not surprised. I would have flunked it too.

What's in a grade? I have spent my adult life grading college biology students' work. Using tests and projects and class presentations, I can determine which students deserve A grades and which do not. But how to grade a lake? The map maker's criterion is Secchi depth visibility. The clearer a lake's water, the higher its grade. Realtors I've spoken with concur.

A chart on the map's backside ranked the ten clearest lakes and the ten murkiest of those sampled. Of 145 lakes, only one had more degraded water than Diamond. No other lakes within a dozen miles had been sampled, leaving the two red dots alone at the north end of the county, like unruly schoolboys banished to the corner of the room in punishment for egregious misbehavior, isolated so as not to contaminate others.

I thought of students who had stumbled badly in their first exam but then made changes, overcame circumstances, and improved their performances dramatically, in some cases moving from Fs to As. Is Diamond condemned forever to be limnology's equivalent of the dunce-headed schoolboy? Is there any hope that if circumstances changed it might someday overcome its shortcomings? Can Diamond change its dot?

At first glance, dot colors seemed randomly scattered across the map. As my eyes moved more deliberately, however, a pattern emerged. A dense aggregation of dark-blue dots, A lakes, were congregated in the long, narrow easternmost county. I counted them. Two-thirds of the map's twenty dark-blue dots lay clustered in a space less than a tenth the area of the entire map. Diamond Lake's county, two and a half times as large as the blue-dot-studded space, contained not a single such dot.

Patterns. Sometimes they are mere artifacts of random chance signifying nothing. Or they may reveal clues to causes and relationships. I began wondering if blue dot lakes might yield insights as to what led to Diamond's failing grade and might hint at its prospects for a brighter, clearer future.

On a sunny July day Geri and I set out to explore a blue-dot lake to find out. A beautiful, rolling glacial moraine countryside unfolded as we approached the lake. Big Carnelian sits on the southern edge of the blue-dot lake cluster. Its Secchi depth of 12.8 feet ranks it sixth clearest of the 145 lakes sampled. The chart also reports Big Carnelian has one-seventh the total phosphorus concentration as water in Diamond Lake. At 455 acres, it is only slightly larger than Diamond.

I arrived at the lake's public landing with a suspicion. Excess phosphorus, leached by runoff from well-fertilized cornfields and residential lawns and years of seepage from septic systems on the north end of Diamond Lake surely caused much of that lake's murkiness. Land use around these dark-blue-dot lakes must contribute much less phosphorus to lakes like Big

Carnelian. They must have more pristine, less disturbed shorelines. At least that is what I expected.

As we set the canoe in the water, a man with a small kayak joined us at the dock. He lived only a few miles away and said he had paddled Big Carnelian and other lakes in the area before. "Are you a fisherman?" I asked.

"No," he replied. "I find lake creatures fascinating. I come to look and take pictures."

"Oh, you're a biologist," I replied, sensing a kindred spirit.

"Nope, I sell real estate. I'd like to own a place on this lake someday."

We wished each other well and the man began poking his way north into a dead-end bay rich with plants. Geri and I paddled south, sauntering, as Thoreau might put it, slowly along the shore. Sauntering a lake perimeter is much like walking a city neighborhood. I come away from such poke-along travels with a much deeper sense of people, how they live, and what they hold dear.

My first view of Big Carnelian was not what I expected. The near-shore surface of this small bay was rife with mats of algae and aquatic plants. Houses and lawns nestled in long rows along the beach seemed a likely source of the nutrients responsible for the profusion of plant life. Surprisingly, open water patches scattered among the mats were pleasantly clear.

I jotted down field notes along the way:

So many cabins and houses on the shore, few stretches of undeveloped property.... Strong winds rile the bottom on the east shore. Hard to keep canoe from bumping into docks, tethered boats and floating diving platforms.... Now in small protected bay— densest collection of boats, docks and cabins I've seen in years. Fifty foot lots with small cabins right at edge of the sandy beach. Shoreline development has been here a long time. Modern zoning rules designed to protect lake water quality would never allow such developments now....At the last of the narrow lots immense pile of Chara drying on dock—another pile lies on the shore—large section of lake bottom cleared of it.

The owner who did the clearing may not have known about *Chara*, the plant-sized alga I saw snorkeling in Eden, before he raked so much away. It hugs the bottom, keeping waves from riling up bottom mud and resuspending silt and phosphorus.

Four young boys rocket themselves skyward off a giant inner tube trampoline anchored beyond a dock. Arms and legs and shouts and laughs fill the air as they careen over the side, splash into the water upside down, then climb aboard for more.

I wanted to join them. What carefree fun unencumbered by concerns about lakes' futures.

Five new houses spread immaculate carpet lawns side to side and from stoop to sandy shore. A thousand feet of golf course–like savanna, nothing to interrupt the lawn grass landscape of suburbs. . . . NW shore: Mesh holds plantings of sedge and other shore plants at the upper edge of someone's beach. Next two lots the same thing. Owners trying to protect the lake from the land with a buffer belt of plants. . . . Next lot down, a narrow grassy path leads to a dock. Rest of shoreline undisturbed as can be. Those folks love their lake.

How different are the lakescapes in different minds.

Still NW shore—Chara covers much of the bottom several feet below water surface. Much greater variety of pondweeds and other aquatic plants than on Diamond. Grand starter castle estate dominates a point. Tightly mowed grass to the water's edge across an immense yard. Nothing to hold back runoff and leachate from the land. "Look at me," the castle boasts, "Look at me."

Homes and cabins were much denser than I expected. Only a small percentage of the shore is undeveloped. Home owners typically fertilize lawns at much higher concentrations per acre than do farmers. Nutrient runoff into the lake is unavoidable. How can the water in this lake be so much clearer than in Diamond? Strong winds prevented me from getting a Secchi depth reading.

A swan family snoozed on the shore of a tiny island as we passed. Two adults, three cygnets—priceless. I understood no more as we approached the landing than I did when we began. Maybe a different blue-dot lake holds clues.

·   ·   ·

A month later I shoulder the canoe and set off alone down a narrow wooded trail toward another of the blue-dot lakes. Low tree branches scrape and screech across the canoe as I follow the path's twists and turns through dense buckthorn brush and small oak and birch trees. A green chain-link fence separates the public land I walk on from private property. It's August. Plant and algal growth should be at their peak and lake water at its least clear. The path finally ends at a narrow sandy beach, the lake glistening in bright late morning sun. Less than half the size of Diamond, this is no ordinary blue-dot lake. It ranks at the top of the As, valedictorian, as it were, of last year's class, the best of the best, clearest of the clear. Little Carnelian lies less than a mile from its bigger brother to the north.

I flip the canoe off my back and set it in water so clear I can distinguish individual sand grains at unbelievable depth. I load my lunch, snorkel gear, Secchi disc, and paddle and shove off. A light breeze pushes small waves at me from across the lake.

Down the beach a young man casts a fish lure far out into the water, aiming at the edge of a thin band of pondweed that parallels the shore. "Any luck?" I yell.

"Couple of bass," comes the reply. "Nothin' big."

The ends of sawed-off tree trunks project out of crystal-blue depths to within several feet of the lake surface. A tan encrustation envelopes these submerged boles. A poke of the paddle dislodges a small piece. It is firm. It might be a freshwater sponge. Sponges are very persnickety about where they live, demanding very high water quality. Most freshwater sponges I've seen are green, however, so this could be something else.

I swing the canoe toward the center of the lake and lower the Secchi disc into the water. The plate's white quadrants finally disappear twenty-four feet down. That's nearly twice the visible depth reported for Big Carnelian, and twenty-four times the disc readings I get at Diamond Lake.

I return to shore and look into the exquisitely clear water. *Chara* dominates the bottom at depths of one to four feet. Deeper, the tall cylindrical columns of native milfoil and patches of broad-leaved and fern-leaved pondweed replace the *Chara*. Fishes dart out from the vegetation to greet my approaching canoe.

I exchange paddle for snorkel and glide magically through an underwater pondweed forest, a fairyland of fishes and spindly plants eight feet tall. Skeletons of submerged tree branches create shadow and mystery and havens in which to hide. It hurts to leave this enchanting kingdom for the surface and a breath of air.

A narrow finger bay in the northwest corner of the lake lures me in. It curves like a hook, disappearing from view from the rest of the lake. I follow it and am soon in stagnant water, then muck. I pull the paddle hard to penetrate a thick green mat of filamentous green algae and duckweed. An unpleasant stench of rot displaces the alluring smell of the open lake.

The channel ends at the mouth of a large culvert. I have stumbled on Little Carnelian's nasty secret. Obviously the culvert, the end of a storm sewer pipe, delivers some not-so-welcome silt and chemicals to this hook arm. What a contrast to the main lake. Today the arm acts as a sediment and nutrient trap protecting the water of the main lake. There has been no rain for days. I can imagine what could happen when a gully-washer arrives.

Back on the main lake I rest my paddle and take a deep breath. Though the shoreline here is not undisturbed, this lake feels more relaxed than Big Carnelian. All told, a dozen or so well-spaced houses are interspersed along a mostly wooded shore. Some have full-width lawns, though not the dense, heavily fertilized carpet lawns I saw in places on Big Carnelian. Houses and cabins do not overwhelm the shore. The relationship between lake and land feels less frenzied, more subdued. I think I understand how this lake remains clearer than its brother to the north. But how does either remain so much clearer than Diamond?

Fish swim lazily within a pondweed forest deep beneath my canoe. How exquisitely clear the water. Every lake lover should go on a pilgrimage a least once to experience a place like this, to see what a lake can actually be.

A schedule calls. I head toward the landing. Midway down the shore, the sand becomes progressively more gravelly. I now notice scattered red pebbles mixed among those of tan and black and gray. Diamond Lake also has several gravelly shores, but I don't recall seeing red pebbles there.

As I carry the canoe back up the winding path to the car, my mind returns to the red stones. Red pebbles here. None at Diamond Lake. Suddenly my mind is seized! Two glacial lobes—from different directions—carrying different kinds of till! Why hadn't I thought of it before?

Bits of red sandstone are the hallmarks of an ice lobe that entered the state from the northeast, from Lake Superior way, scraping over ancient red sandstone as it passed.

The other major lobe scraped across parts of Manitoba and eastern North Dakota. It carried clay and much limestone but *no red pebbles.* Both the Superior Lobe and the Des Moines lobe from the northwest left till in this part of Minnesota. Absence of red pebbles at Diamond means its surroundings are composed of different minerals than the landscape of the Carnelians. What a tantalizing thought: different till, different chemical makeup, different lake chemistries. Different lake chemistries yield different lake behavior. I find the logic compelling.

My map at home of surface glacial deposits of the state is unequivocal. Diamond Lake lies in Des Moines lobe till and the blue-dot lakes of northern Washington County lie in Superior Lobe till. So that's how the Carnelians and Diamond came to such different fates—or so it seems until I reexamine the report card map and discover something I'd previously overlooked—a bright-red dot sitting fat and sassy among the A lakes no more than a mile, as the swan flies, from Little Carnelian. Loon Lake, an F lake in Superior Lobe till. My hypothesis withers.

Maybe Loon had some quirky origin, I thought. Maybe a geological aberration had isolated it from the surrounding Superior till making it behave like Diamond. I went to see the lake for myself. Wish as I might, the gravelly shore of Loon Lake provided no reprieve for my ill-fated hypothesis. There they were, those pesky red pebbles, just as they were in the Carnelian lakes. Loon Lake, with its F grade, sat in the same type of till as its blue-dot neighbor.

I retreated to the official monitoring records for the clearest A and murkiest F lakes on the report card map, seeking other explanations of why some lakes remain clear when others become green soup. There I discovered the F lakes share two things in common. All are shallow, less than fifteen feet deep. Diamond is seven feet at its deepest. All of the A lakes are deeper than that, most are significantly so.

Secondly, the F lakes drain substantially larger watershed areas than do the A lakes, in proportion to lake size. The greater its relative drainage area, the greater the opportunity for unwanted chemicals to get carried into the lake. Loon Lake has half again as large a watershed-to-lake surface ratio as the Carnelians. So, Mother Nature, through geological events long ago, had stacked the deck against Loon and Diamond and had created their vulnerability: shallow water and a large watershed.

· · ·

Diamond, the Carnelians, and most of the other lakes on my map originated when orphaned glacial ice blocks, abandoned in moraines and till fields, melted away. Who knows, had its ice block been bigger and deeper, Diamond might today be a bright-blue dot on my map, with water as clear as the Carnelians.

. . .

I might have laid the issue to rest, leaving my mind cradling such wistful longing, had I not made a serendipitous visit to a library looking for something quite unrelated and stumbled on an account of a Wisconsin lake that had suffered a fate similar to Diamond: high turbidity, loss of aquatic plants, and low Secchi visibility. One difference: more than a decade ago, Milltown, Wisconsin, and the state's Department of Natural Resources tried to reclaim their lake after a history of abuse. Had it worked? My search instantly sparked back to life. I checked out *Wisconsin DNR Technical Bulletin No. 186* from the library and buried myself in Milltown's Rice Lake story.

Historical records document that Rice Lake, a small body of water, had abundant aquatic plants in the 1950s and 1960s, wild rice, flocks of ducks, water so clear boaters could watch fish cruise below, and kids swimming. Then conditions deteriorated. Muskrats and water level changes reduced rice, so ducks diminished and algae and bullheads reduced water clarity. People had quit swimming and fishing on the lake by the early 1970s.

The report surmised that a combination of decades of nutrient input washing downstream from the town and surrounding farmland, in combination with unusually high water levels and bullheads stirring up bottom mud, had ultimately eliminated most of the aquatic plants, including wild rice. "Loss of wild rice from the lake and creek, after decades of water quality change, proved crucial: cover for waterfowl was destroyed, [insect and microcrustacean] prey for panfish declined, and summer winds scoured the bottom. . . . Water turbidity created by algae and suspended particles limited aquatic plant growth." In a word, the lake had flipped from being a stable clear water lake with plentiful plants to a murky algae-dominated, windswept lake, with greatly reduced aquatic plants.

Elements of Milltown lake's story paralleled what I'd seen on Diamond. Increased nutrient inflow stimulated plant growth to nuisance levels.

Continued nutrient input increased algae numbers and supported a thickened layer of periphyton on plant stems and leaves, robbing plants of light.

In the last few years the formerly massive mats of elodea and coontail
on Diamond have dramatically receded. Diamond itself has progressively
flipped over to an algae-dominated condition. The demise of aquatic plants
removed a refuge for creatures like daphnia that had been reducing algal
numbers through their predation. Fewer plants left more phosphorus for
blue-green and algae populations to expand further.

With rooted plants no longer stabilizing bottom sediments, winds stir
up mud, adding to the lake's turbidity, making it even less likely plants can
recover on their own. A three-mile-per-hour wind will stir up over half the
lake bottom of a lake of Diamond's depth.

The key to recovery of Milltown's lake, limnologists felt, was the return
of rice and other aquatic plants. More plants mean less resuspension of
bottom mud, and enhanced diversion of phosphorus to plants and away
from cyanobacteria and algae. Restoring plants would be key to returning
the lake to a clearer, more aesthetic condition.

Restoration began in the late 1980s and early 1990s. People planted
tubers of wild celery and wild rice into Milltown's lake. Now, nearly twenty
years later, what had been the result? I had to find out. Geri and I drove to
Wisconsin and stopped at a hardware store in Milltown to get directions to
the lake. The first two people I asked didn't know such a lake existed. A third
man did and offered to lead us the two miles out of town to the boat landing.
Pop and beer cans and bits of paper and plastic lay scattered around a weed-
choked fire ring of blackened rocks. No one was on the lake, and the landing
gave no evidence of recent use.

My report on the lake's condition is bittersweet. The lake indeed has
flipped from murky, algae-dominated with few plants back to the clear
water, aquatic plant state. Its water is much clearer than Diamond's, and
algae and blue-greens no longer dominate this lake.

Though I could find no rice, I did find a rich diversity of plant life.
Sago and floating-leaved pondweed and wild celery, species of particular
importance to waterfowl, abound. I found *Chara* (Even Geri, sitting at the
other end of the canoe, could smell its tell-tale musky skunklike fragrance),
elodea, lilies, and more. I even found sunfish nests.

But the lake's flip back to a clearer water condition may have come too late. I could not get an accurate water clarity reading. My disc disappeared into a black, fluffy layer of gyttja a little over four feet down. The many decades of nutrient flush from the city's mills and farmlands in the watershed apparently stimulated growth that sped the process of lake fill-in.

So the lake flip has succeeded in one sense and failed in another. In such a shallow condition, the lake appears all but unrecoverable for most recreational uses, despite its return to clean water.

.   .   .

Engineering a lake's flip from being highly turbid and algae-dominated to having clear water and abundant plants is difficult. Every lake is unique. What works for one may not work for others. Paradoxically, dramatically reducing the phosphorus now entering a lake, the culprit that set the lake's deterioration in motion, likely would be ineffective. Too much phosphorus already resides in the sediments and cycles up from bottom muds. Loss of the aquatic plants dramatically rejiggered the entire system.

.   .   .

The final turn in my search for the hard bottom of truth about Diamond Lake and prospects for its future came when I attended the annual meeting of a lake association as my father-in-law's guest. An older man who owns a home on the lake approached me with a question. "A large weed bed forms every summer in front of my house. Which chemical is the best to clear that problem away?" He went on to describe the problem plant, and I realized it was not one of the troublesome exotic species.

"Having problems getting your boat through it?" I asked.

"No," came the reply. "I don't go out on the lake. I've got no interest in fishing or skiing, that sort of thing."

"So why do you want the plants removed?"

"I just like to look at the water. I don't want to see weeds."

I couldn't help overhearing another conversation two tables away as neighbors, in animated voices, shared stories of fish caught the past summer at the edges of *just such weed beds.*

My mind grew uneasy. Had my perceptions as a scientist left something out? Were the stepping stones to truth those of lake basin chemistry, mineralogy of glacial till, Secchi depths, micrograms of phosphorus and the like, as I had believed? Or, is lakescape as perceived by the mind a more direct path to the hard bottom of truth? Do human perceptions trump scientific truth?

Does Diamond's future depend less on the actions of long-ago glaciers and more on our perceptions of lakes and what we want them to be? Would Diamond Lake's users even welcome a flip from a blue-green dominated lake to one dominated by plants, with its inconveniences for water-skiers and others who value freedom of speed? A bizarre image forms in my mind. I see one group of people transplanting aquatic plants into one end of a lake to enhance water clarity and fish habitat, while others remove plants at the other end.

Limnos IV—Cedar Bog Lake, Minnesota

The organisms within an ecosystem may be grouped into a series of more or less discrete [feeding] levels, producers, primary consumers, secondary consumers, etc., each successively dependent upon the preceding level as a source of energy, with the producers directly dependent on solar radiation as a source of energy.

RAY LINDEMAN, 1942

John turns the car onto a two-rut sand trail. We drive across a field, stop at the edge of a woods, and get out. A signboard welcomes us to Cedar Creek Bog, part of the University of Minnesota's Cedar Creek Ecosystem Research Center. The signboard explains that the irregular knob and kettle topography of the area resulted from the melting of a large block of glacial ice amidst glacial outwash sands.

John Haarstad, staff member at the station, has agreed to take me on a walk this morning to a small lake within the bog that holds particular interest for me. Seventy years ago a young doctoral student came here to study the inner workings of the aquatic community of that tiny body of water. He left with insights that transformed ecologists' perceptual understandings and propelled Cedar Bog Lake into ecological history. For me, the man's personal story is inseparable from this lake and the discoveries he made there.

I first learned about Ray Lindeman when, as a beginning graduate student at the University of Minnesota, I was assigned office space in Room S2 in the basement of the Zoology Building with other ecology graduate students. I heard officemates talking in casual conversation about someone named Lindeman. I asked about him. "He was a grad student of Sam Eddy's," came the reply. "Did his field work up at the Cedar Creek Station."

"Lindeman made quite a splash in the field of ecology," another said. "Did you know his office was in this room?" Over time I came to understand why he was spoken about with a sense of awe.

• • •

A boardwalk leads away from the sign into an oak and maple upland. The woods' understory is open, the walk cool and pleasant. A few feet drop in elevation and the trail passes into a cedar bog and then a black ash seepage swamp. "We call this Crone's Island," John says, as we walk up a small hill back into oaks and maples. "This would have been an island in the original glacial lake."

Occasional white signs the size of large postcards identify plant species along the boardwalk. John's doing. We stop so he can take pictures of a tiny rare plant not yet acknowledged by a sign. John identifies other plants as we walk and tells me of his own graduate studies of dragonflies on Cedar Bog Lake. The boardwalk is slippery from last night's rain. We watch our steps. Sooner than I expect we see joe-pye weed, boneset, and a sagittarian, plants that prefer wet feet. We are getting close.

Now the trail breaks into the open beside the lake. We step from the boardwalk onto an old dock mostly buried in peat, then onto a newer higher dock that extends closer to the water's edge. The lake is shaped like a tear drop and seems no more than two football fields long and half that wide. John points out the swamp loosestrife that rings the open water around the lake's perimeter. Scattered cattails, alder brush, tamarack, and clusters of poison oak appear farther back from the water. I take a step off the dock onto the bog mat. It wiggles and sinks beneath my foot, unable to support my weight.

A dense stand of coontail covers much of the lake. John says he's seen it even more widespread. "At times you feel with big enough snowshoes you could walk across from shore to shore. When that stuff decomposes it must really speed up the fill-in of the lake." John becomes quiet. I look out over the senescent lake. What a humble birthplace for an idea that would become a foundation of modern ecology.

Lindeman wrote that in his time the edge of the mat was encroaching out onto open water nearly a meter every five years. "Do you know how much the open water has been reduced since Lindeman was here?" I ask.

"I don't, but this semi-floating dock was put in last year right up to the edge of open water. Look at it now." I look down. Swamp loosestrife has extended itself almost a meter farther out into open water in just one year. Lindeman recorded that sediment cores taken in his time revealed that deposits of gyttja and marl had filled in close to 90 percent of the lake's original depth. This lake is dying.

· · ·

I think about Lindeman and the number of times he would have walked the path, packing or dragging in equipment, and all the rowing, dredging, plankton net towing, measuring, and weighing during his five years of work here. Already handicapped by blindness in his right eye, Lindeman developed a liver illness. As his health worsened, his wife Eleanor's help with the field work became essential. At times he vomited blood and became hospitalized for days. Doctors could not identify what was wrong. As soon as he was able, Ray and Eleanor would return to the lake to resume the research. Despite his deteriorating health, or maybe because of it, he threw himself more intensely into his work, not even stopping for supper unless Eleanor was there to demand it.

Lindeman studied all the elements of the Cedar Bog Lake community from plants and algae to plankton and insects living in the bottom mud. Then he took a step no one before had taken. He translated all their feeding interactions into a single parameter—the movement of energy through the feeding levels. He saw the lake as a union of living and nonliving components creating a unitary functional system, an ecosystem.

The same Professor Eddy whose class field trip had drawn me to biology had been Lindeman's adviser. The scuttlebutt in Room S2 among my fellow graduate students was that Eddy didn't truly understand the significance of his student's insight.

Lindeman received his doctorate from the University of Minnesota in March 1941, and in August he left for Yale to take a postdoctoral position under the tutelage of the eminent limnologist G. E. Hutchinson. In October,

Lindeman sent off the final, more theoretical chapter of his doctoral dissertation to the journal *Ecology* for publication. A month later he received a rejection from the editor. The editor had sent the manuscript to two leading limnologists in the country for their review. Paul Welch at Michigan questioned the suitability of Lindeman's work, writing, "It seems to me unfortunate if the space which should be occupied by research papers is partly consumed by 'desk produced' papers. . . . The basic background data for such a paper is far too fragmentary. . . . Limnology needs . . . research of the type which actually yields significant data rather than postulations and theoretical treatments."

Chancy Juday at Wisconsin concurred. "A large percentage of the discussion and argument is based on 'belief, probability, possibility, assumption and imaginary lakes' rather than on observation and data. The chances are that the author's beliefs and imaginary lakes would be very different entities if he had a background of observations on fifty or a hundred . . . lakes . . . instead of only one." Sam Eddy was not the only one to fail to understand Lindeman's genius. Perceptual change comes hard.

After receiving additional advice, the courageous young editor of *Ecology*, Dr. Thomas Park, informed Lindeman that if he made at least some revisions he would publish the paper, despite the considerable opposition.

Lindeman's health did not improve. In April he wrote to a close friend: "We hope to be at the University of Pennsylvania next year, as I have a fellowship there, but (confidentially) there is a better than even chance I don't survive the summer. My liver trouble has gotten irregularly worse, in spite of the best doctors." In mid-June doctors performed exploratory surgery. Two weeks later, on my first birthday, Lindeman died. Since his paper was still in press, he never saw it in published form.

Sadly, the twenty-six-year-old never lived to hear the comments of others on his work: *pathbreaker . . . the most significant formulation in the development of modern ecology . . . . Lindeman's article opened up new directions for the analysis of the functioning of ecosystems . . . . one of the most creative and generous minds yet to devote itself to ecological science . . . . Ecosystem science was born at Cedar Bog Lake . . . . Ecology in 1940 lacked a validated conceptual framework . . . [the] young man was to change all this during what became, at his personal level, an agonizing race against the Grim Reaper.*

Lindeman's insights set the stage for ecologist Gene Odum to popular-
ize the notion of the ecosystem in the 1970s as an integrated functional
unit, as something greater than the sum of its parts. But if everything is
connected to everything else, then everything is mutually dependent. Since
humans are part of ecosystems, what humans do to them matters. Odum's
formulation provided great impetus to the environmental movement but
raised the hackles of individualists. Alston Chase lamented, "If everything
is dependent on everything else . . . then all living things are of equal worth,
and the health of the whole—the ecosystem—takes precedence over the
needs and interests of individuals." Indeed, holism and ecological reality
have consequences, even for human behavior.

.   .   .

John and I turn to the boardwalk and begin walking back to the car. I ask.
"Do many people come to visit Cedar Bog Lake?"

"Oh, occasionally some come just to see the lake. A few years ago a
couple of Brazilian limnologists came. One of them said as he left, 'Now
that I've seen where Ray Lindeman did his work, my life is complete.'"

Thinking Like a Tullibee

*Meaning is there to be discovered in the landscape if only we know
how to attend to it. Every feature, then, is a potential clue, a key to
meaning rather than a vehicle for carrying it.*

TIM INGOLD

O-do-nee-bee. Cisco. Tullapy. Lake herring. *Coregonus artedii.* By any of
its names, the tullibee swims all but unknown in many northern lakes.
Why this fish is unfamiliar to many people, even fishermen, is no mystery.
Tullibee cruise the open water beyond the shore, in a world seemingly
unconnected to our own. I discovered otherwise in a personal journey that
began one June morning on the shore of Lake Itasca in northern Minnesota.

I had just arrived at the University of Minnesota's Itasca Field Station as a beginning graduate student, eager to begin my thesis research in
lake studies. My adviser, Professor Underhill, had shown me around the
grounds, and we finally arrived at the lake's edge. "Darby, you and John are
to set up offices in the boathouse. Chemical analysis equipment is in the
limnology building just behind us."

As he spoke, I noticed a dead fish with bloated belly floating amid
rushes a few feet offshore. "So that's where the ripe odor is coming from," I
observed, pointing to the faded rotting perch.

"Ripe odor?" came the reply. "Just wait until the tullibee die off. That's
when the smell of rotting fish *really* gets rank."

"Tullibee die-off? What's the story?" I asked.

"Every few years tullibee in this lake die, sometimes by the hundreds," Underhill explained.

"Do we know why?"

"The worst die-offs happen during the hottest summers, so temperature is involved, but probably other factors as well."

Excessive heat. Not a surprising connection. Temperature too high makes chemistry go wrong. Seriously wrong chemistry and life stops. But temperature too high in a Minnesota lake? No swimmer I knew ever complained of such a thing, even in midsummer.

My orientation tour finished, Underhill returned to the lab, leaving me alone. In those moments tullibee, and the uncertainty surrounding their misfortune, intrigued me. I spent the next three years of my life getting as close to their lives and their tragedy as my land-lubber physiology would allow, seeking the cause of their troubles.

. . .

Tullibee. I first met them in ichthyology class. They inhabit cooler northern lakes and are found across much of Canada and north of a line from central Minnesota across southern Wisconsin and Michigan through northern Indiana. Unlike many fish that are at home in both lakes and streams, tullibee are truly fishes of the open lake. Except for a quick sortie into the shallows to spawn in late fall, adults typically stay out from shore. Tullibee's preference for daphnia and other planktonic animal life means they have little interest in the menagerie of lures that inhabit most anglers' tackle boxes, although people catch them through the ice in winter with tiny hooks and tiny baits.

Scattered groups of wooden buildings, all painted state park brown, make up the Itasca Field Station. Classroom buildings with cement floors and long tables sit among pine trees, a line of faculty cabins not far away. Student cabins form a line along the lake on either side of a mess hall. I was assigned, along with several other graduate students, to Cabin 9. A bunk bed stood in each corner of the open room. Earlier arrivals had already thrown sleeping bags and duffel onto thin mattresses to claim a space. A small bare table stood on each wall. Other tables with gooseneck lamps occupied the middle of the room. A single switch at the door controlled a lone overhead light. The bathroom building, shared by residents of all the men's cabins, sat behind cabin row. Though austere, the accommodations drew few complaints. For most graduate students, living arrangements were trivial considerations alongside the stimulating ideas and research questions buzzing in our brains.

A research plan began forming in my mind. While tullibee deaths appeared strongly linked with temperature, other factors—oxygen in particular—might be at work as well. Oxygen is as critical to fish as fishermen, and I understood the difficulties aquatic animals face in obtaining it.

We land creatures can take our oxygen supply for granted unless climbing tall peaks or suffering ill health. Oxygen comprises 20 percent of the atmosphere and is reliably available. Such is not the case for fish. Oxygen in lakes is at a hundredfold lower concentration than in air under the best of conditions. What's more, the supply is unreliable. The quantity of oxygen available can change with season, depth, and even time of day. While the correlation with hot summers linked the die-off directly to water temperature, the cause of death might involve a temperature-oxygen interaction.

I decided to track temperature and oxygen concentration patterns from lake surface to bottom throughout the summer. Then, by suspending a gang of long nets at different depths and recording which depths yielded fish and which did not, I could determine tullibee response to changing water conditions. By knowing the conditions where tullibee lived when the die-off began, I hoped to reconstruct the specific circumstances responsible for their deaths. By the end of my second week I was ready to start sampling Itasca's waters.

I loaded my equipment into a small aluminum fishing boat. An electronic probe on a long black insulated wire coiled around a circular black wheel measured temperature. An aged wooden pop-bottle crate contained the oxygen analysis equipment: water sample bottles, a hollow brass cylinder with rope attached to collect water samples, eye-dropper bottles containing chemicals that would reveal oxygen concentrations, and a notebook. I undid the boat's mooring ropes, pulled the motor to life, and putted away from the dock. The search for answers had begun.

· · ·

Lake Itasca is a thousand-acre body of water composed of three long narrow arms joined together to form the letter "h." Each arm contains its own deep spot. The Mississippi River, originating in the lake, flows north out of the top of the "h." I steered the boat away from shore, south past Schoolcraft Island, rounded Bear Paw Point, and headed toward the Peace Pipe Basin at the lake's southeast arm. Peace Pipe, the deepest spot in the lake, seemed the most likely refuge for stressed tullibee.

As the boat glided to rest, I dropped anchor, unwound the tele-thermometer, opened my notebook, entered June 20, and lowered the probe into the water. Surface temperature seventy degrees. Temperature three feet down, seventy degrees. Six feet down, again seventy. The thermometer seemed stuck on seventy as I lowered the probe further, until reaching the sixteen-foot mark. There, almost instantaneously, the temperature dropped nine degrees. The sharp temperature drop reminded me of surface-diving as a kid. Swimming through the temperature gradient like the one I had just recorded had been memorable, like swimming through an open refrigerator door. The next deeper measurements were cooler, and the temperature continued dropping, but at a much slower rate, until bottom, where the probe reported forty-eight degrees.

I lowered my water sampler, an ingenious brass cylinder that allowed me to capture water samples at whatever depth I wished.

My first sample came from just below the lake surface. I transferred the water to a bottle where measured squirts of several chemicals reacted with the sample's oxygen, forming a billowing tan cloud. After shaking, a thick brown flocculent material settled into the bottom half of the bottle. A squirt of acid dissolved the floc into a clear deep amber solution that signified oxygen aplenty. Back in the lab I could precisely measure the quantity of amber molecules. More deep amber bottles from successively greater depth joined the first.

Then, with a sample from halfway to the lake bottom, the color intensity began to fade, with deeper samples successively lighter shades of yellow. The lighter the color the less oxygen the sample had contained. The last bottle, the one from the lake bottom, was clear. Though lab work would assign precise oxygen concentrations to each, the overall pattern was obvious. The lake water was layered—warmer water with plentiful oxygen near

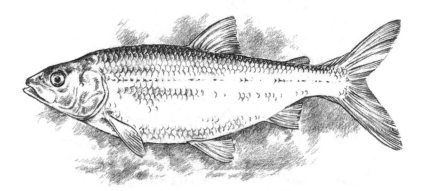

Tullibee originally ranged from northern Indiana across much of Canada, but southern populations are now seriously stressed.

the surface, a deeper transition layer where conditions changed sharply, and colder, poorly oxygenated waters at the bottom.

Next, I had to find the tullibee. Underhill arranged with the Department of Natural Resources to loan me six gillnets 250 feet long and 6 feet wide. I would suspend the nets at different depths by attaching floats with different length ropes for each net. By comparing the size of catches from each net over time, I expected to track changes in the vertical distribution patterns of the fish.

The day of my first set I scurried to gather washtubs to hold the nets and outfitted floats with proper length ropes. On reaching Peace Pipe I attached an anchor to the end of the first net. I had no experience with large nets, and the challenge of my task suddenly became clear. Setting a taut net required me to keep the motor running at just the right speed and the boat traveling in a straight line—while simultaneously uncoiling net from the tub and attaching float ropes at equal intervals to suspend the net. Run the motor too slowly and the net lacked the necessary tension. Run it too fast and hands couldn't get float ropes properly attached before sections of the net got yanked overboard.

I also had not anticipated the uncooperative behavior of the nets themselves. They became tangled for no reason at all, seemingly in spite, forcing me to endlessly cut the motor, to endlessly untangle the obnoxious mesh. I simply had not enough hands. I finally reached the end of the net, attached the anchor, and dropped it over the side. I looked back at the set. Floats bobbed in chaotic disarray. Far from the taut straight line required, the net hung in loose folds, like a massive disheveled cloak heaved overboard. I killed the motor, jerked the uncooperative net back into the boat, and dumped it into its tub armful by armful. Pulling the nets would present none of these problems, but setting them would require help.

Fortunately, other grad students had their own needs for assistance, and by the next day John and I had a mutual-aid agreement. He would run my motor while I set nets, and I would help him screen his mud samples for bloodworms, phantom midges, and other life at the bottom. With John at the throttle and me at the nets, the setting went smooth as clockwork.

Next came the wait. Science is a delightful world for the curious. Fascinating, compelling questions abound. The answers, however, are like cryptically wrapped Christmas presents deviously hidden from view. I dreamed of fish and nets all night.

Next morning the boat sped me back to Peace Pipe, and I pulled the surface net first. Empty. Two silvery fish wiggled in the nylon webbing of the next deeper net. I had seen tullibee pickled in formaldehyde in ichthyology class, and in fish markets, stained dark brown from being smoked. Finally, the chance to see them in life. Pink iridescence glistened off their silvery sides in the sunlight. They were beautiful.

Identification was easy. Small projections at the base of the pelvic fins and a tiny odd rubbery projection on their backs, in front of the tail, placed them in the Family Salmonidae. Absence of robust teeth distinguished them from their cousins the salmon and trout. Lack of an underslung mouth distinguished them from closer cousins, the whitefish. Their small, dainty mouths suggested personalities quiet and reserved, in sharp contrast to the gaping maws of walleye, bass, and pike bristling with teeth.

I pulled the next deeper net. Empty. Then bonanza! The next two nets held dozens of glistening tullibee. Most ran about a pound and a half. Not a game fish among them. I continued pulling the nets and recording the catch.

Back at the boathouse, with oxygen, temperature, and tullibee numbers transferred onto a graph, the situation became clear. From the lake surface down to seventeen feet only two tullibee had been caught. Nearly ninety fish came from the layer of water between seventeen and twenty-nine feet. Water below twenty-nine feet yielded only six fish. The lake had three layers of water, like a three-level house. The upper level, the upstairs, contained warmer, well-oxygenated water. The main floor, the middle zone, was thirteen feet thick where 90 percent of the fishes had been caught. The bottom layer of cooler, less oxygenated water formed the basement.

A quick graph revealed that the middle zone the fish inhabited had a ceiling temperature of sixty-eight degrees, and a two parts per million oxygen boundary formed its floor. The fish inhabited water cooler than sixty-eight degrees that contained two parts per million oxygen or more. But were these fish in the main floor because they sought the temperature and oxygen levels there or because of something else? A school of tullibee, temporarily swimming in the seventeen to twenty-nine-foot stratum, might have blundered by blind misfortune into my nets, while other schools at different depths chanced to miss them. My suspicions about tullibee depth preferences necessarily remained tentative, pending results of future nettings.

. . .

Days later I repeated the netting and oxygen and temperature measurements under beating sun. The new numbers resembled my earlier discoveries, but conditions had deteriorated for the tullibee. The sweltering July heat had further warmed the entire upper layer of water. Surface temperature now stood at seventy-seven degrees. The sixty-eight-degree temperature ceiling of the tullibee's main floor home had lowered a full three feet.

Oxygen conditions in the basement had worsened as well. Oxygen was totally absent from all waters deeper than thirty feet. The basement had expanded upward, pushing the two part per million floor a full three feet higher. With its ceiling lowered and its floor raised, the tullibee's living space had become scrunched. I caught not a single fish upstairs above the sixty-eight-degree temperature line nor below the two parts per million oxygen line in the basement. Now all had crowded into the diminishing space of the main floor.

Imagine their distress. To avoid oppressive temperatures threatening to throw their metabolism into chaos, their instincts must surely be to dive deeper, to get to the cooler, more comfortable basement. But there was no oxygen in the basement. To go there courted asphyxiation. The image of mouths gulping and gill covers flailing frantically in a desperate race to match oxygen intake with body demands distressed me. In a cruel twist of fate for animals like fish, where body temperature takes on that of the external environment, a rise in body temperature, all by itself, increases the need for oxygen.

During the next net set, a storm blew up during the night, tearing floats loose and wrenching nets into twisted, sagging masses. Instead of recovering tullibee that day, I spent two hours in the boat removing dozens of tiny perch from the webbing, the fishes' sharp dorsal spines treating my fingers like pin cushions.

New temperature and oxygen data solidified my suspicions. The upstairs and the basement both were expanding into the precious space of the middle zone where the tullibee lived.

The increasing water temperature in the upper waters of the lake was the predictable outcome of hot air temperatures. But what explained the rapid disappearance of oxygen at the bottom? Obviously more oxygen was being consumed than was being replaced. But fish like tullibee were no longer present in those depths to consume it.

Birge and Juday had explained the phenomenon many decades before. Oxygen enters lakes in two ways. Much diffuses into surface waters from the massive oxygen reservoir in the atmosphere. But diffusion through water is terribly slow. Without being physically mixed, it would take years for oxygen to reach the bottom waters on its own. Summer winds mix lake water, but only in the upper layer above the thermocline (that refrigerator door). For fish in deeper waters, surface oxygen might as well be on the moon. Aquatic plants and algae also put oxygen into a lake as a by-product of photosynthesis. But since lake water absorbs light with depth, photosynthesis is limited in deep waters.

Itasca's bottom waters were cut off from oxygen resupply. What caused its disappearance may surprise you as it once did me. It's bacteria, plain, ordinary, decomposition bacteria that thrive on dead bodies for food. If one's palate runs to dead things, a lake bottom is a wonderful place to live.

Lifeless plant, algae, and animal bodies rain down from above, littering lake bottoms with things bacteria like to eat. Many bacteria use oxygen as they decompose dead matter and are so effective they can remove it all from parts of a lake. Lake bottom: graveyard to some, cafeteria to others. The tullibee's plight hangs in the balance of all this dining.

Nutrient-rich waters beget much plant life. Abundant plants feed abundant animals. Many creatures ultimately mean many deaths and a downpour of food to the lake bottom, therefore lots of bacteria slurping oxygen.

· · ·

Underhill and I met weekly to talk over my progress. He taught the limnology class, and, through class studies, he and his students also knew of the temperature and oxygen conditions in the lake. "So, Darby, the heat wave continues. How are those tullibee of yours doing?"

"They're stressed. My netting shows they're avoiding both surface and bottom waters. They've crowded into a middle layer that becomes more compressed with each successive netting."

"From what I've seen in past years, I'd say, unless this hot weather breaks, a tullibee die-off is coming soon," he replied. "You'd like to know precisely when it starts. Why don't you make an announcement at lunch for people to let you know if anyone sees any dead fish."

The first ring of the mess hall's old school bell called "food in ten minutes" across the grounds. At the second gong, the gathered crowd stepped from the adjacent assembly hall up into the dining room with its two rows of long tables separated by a wide aisle. The room soon buzzed with conversation. With the entire station gathered in one place, mealtime was also announcement time. Spoons tinkling against water glasses quieted the din. The station director rose to announce that the fire ecology talk scheduled for that night was postponed to the following evening. The ornithology professor then stood to remind his students to meet in the parking lot at 4:30 the next morning for their field trip. A chorus of groans followed. I stood next. "With all this heat, a tullibee die-off may start soon. If anyone comes on dead fish on the lake, please let me know." I sat down and the room again buzzed with conversation.

In late afternoon an undergraduate student approached me as I was leaving the library. "Aren't you the guy interested in dead fish?" he asked.

"I sure am," I replied. "Seen any?"

"Yes. My girlfriend and I just came back from cooling off at the beach. There was a big one floating by a reed bed not far offshore."

I hurried to the boat and in fifteen minutes arrived at the beach. Sure enough, a dead fish, a big pike, not a tullibee, floated at the surface, a long gash on its side. It must have been hit by a boat propeller. False alarm. The supper bell echoed down the lake as I turned the boat back toward the station.

The next morning, motoring past Bear Paw Point, I watched an osprey drop like a white rock from the sky, hit the water with a splash, and fly off with something that looked like a fish in its talons. But the bird was too far away to tell for sure. A day later, Underhill informed me his class had seen a dead tullibee by the public campground that morning.

The campground was close to Peace Pipe. I spotted the fish lying on its side. As I got closer I saw a fin quiver. I cut the motor and let the boat glide silently up to it. I leaned over the gunwale and reached down to pick it up. It careened sideways and down to escape, but so weakly it got only a few inches below the surface, then floated helplessly back up beside the boat. The fish's gill cover rose feebly, collapsed, then stopped.

I have thought often, in the years since my days at the station, about the sadness many of us display at the death of an individual animal. I note the lengths to which people go to return a beached whale to the sea. I think of the solemnity and tears with which children bury a small bird found dead beneath a living room window. I was particularly moved by the anguish felt by my students in biology lab when confronted with the task of pithing a frog. Their agony was so palpable, I often took the pain on myself and did the deed for them.

Yet an emotional response is muted or absent at actions that kill not single frogs or single fish, but thousands. Students beset with mental pain at the thought of killing one frog in lab look forward to a new shopping center being built on a filled wetland, killing frogs by the thousands and precluding forever their return. I leave such inconsistency for psychologists to explain.

Not far away I spotted several more fish floating. I gathered dead and dying fish to bring back to the boathouse for further study. With the die-off begun, I needed current oxygen and temperature data quickly. I returned to the station, unloaded the fish, and rushed back to Peace Pipe to sample. The temperature probe read eighty-five degrees at the surface, eighty degrees at three feet and seventy-nine degrees at nine and twelve feet. Temperature continued dropping slowly in the deeper water. Not until twenty-three feet down did the temperature finally reach sixty-eight degrees. Oxygen samples from near the surface all showed the familiar amber color signifying oxygen aplenty, although of less intense color than before. Then suddenly the colors changed to very pale yellow and finally, from nineteen feet down, the bottles were totally clear.

A quickly sketched graph revealed the grim truth. Fully 40 percent of the water column lacked detectable oxygen. The sixty-eight-degree ceiling temperature of the tullibee's zone had crashed through the two parts per million oxygen floor. The downward expanding upstairs and upward expanding basement had destroyed the main floor, obliterating the tullibee's refuge. The fishes' choices were horrifying: escape the heat by descending into the asphyxiating water of the basement or escape asphyxiation by ascending into the cooker upstairs. How miserable the poor fish must be.

I set the gang of nets as quickly as I could to learn how tullibee were responding to this calamity. Most fish hovered right at the two parts per million line, seemingly seeking the coolest water possible yet still minimally oxygenated. Pathetically, many came from the warmer water above, seemingly finding the prospect of death by heat preferable to death by asphyxiation. A pitiable few, in a suicidal move, had chosen to descend to sixty-two-degree water, water with oxygen concentrations so low as to be lethal to fish forced into lengthy exposure. Were these fish swimming alternately up into the smothering heat, tolerating it as long as they could to reoxygenate their bodies, then diving into the basement to cool off, forcing themselves to do the fish equivalent of holding their breath? Only the fish knew.

I began a daily boat survey of Itasca's shore for the several week duration of the die-off to see what else I could learn from the dead fish. My observations spawned as many questions as they answered. Ninety-five percent of the dead fish I found came from the Peace Pipe basin. But early in the summer I had caught tullibee in many other places in the lake. It appeared the population had migrated to the deeper Peace Pipe basin as conditions deteriorated.

. . .

One Sunday, desperate for a break from nets and dead fish and sample bottles, I decided to explore Elk Lake, a small lake adjacent to Itasca. No dead tullibee floated anywhere on its surface, yet Underhill had said Elk had a well-established population. Understanding the contrasting absence of die-off in Elk Lake, I reasoned, might confirm my hypotheses about Itasca, but fieldwork had to wait until the following year.

. . .

The next summer I collected netting, oxygen, and temperature data on Elk. While this lake is much smaller than Itasca, it is almost fifty feet deeper. Water sampling showed progressive depletion of oxygen at the lake bottom and warming at the surface, and tullibee avoiding both just as in Itasca. But the zones of excessive temperature and insufficient oxygen did not occupy nearly as large a volume of the water column as on Itasca. Elk's greater depth presented considerably more water to be warmed and considerably more oxygenated water to be depleted. Even when avoiding extreme surface and bottom conditions, the tullibee's main floor refuge was much larger than in Itasca. Elk's greater depth protected its tullibee from the ravages visited on Itasca.

The die-off on Itasca subsided weeks after it began. Continued netting into late summer verified that though many fish had died, many had survived. My final sampling for the season took place the third week of September and revealed profound changes in the lake. The crisp air of autumn had cooled the lake water to a uniform sixty-one degrees to within three feet of the bottom and oxygen concentrations exceeded seven parts per million throughout the water column. The lake had thoroughly mixed, presenting the fish a grand deliverance from the oppressions of summer. My nets revealed that, freed from restraints of high temperature and asphyxiating oxygen, the fish had spread themselves across all depths almost equally.

. . .

A lake watcher could spend an entire autumn staring intently at the lake and never observe the transformational magic that mixes its waters. Imagine sitting in the crispness of fall in a boat on a quiet lake surface. Invisibly the air transmits its coolness to you and to the water of the lake. Cooled water is denser water. And denser water sinks into the less dense depths below. Cooler air cools water as the days of autumn march toward winter. The colder sinking water behaves as a down escalator, moving surface water with its oxygen toward the lake bottom, displacing warmer, lighter bottom waters, forcing them to ascend, as an up escalator, toward the surface. Day after day the escalators run in tandem, one up, the other down, morphing into a sort of Ferris wheel, cycling water from the lake surface to lake bottom, bottom to surface, surface to bottom.

By returning oxygen to the bottom, the circular sinking-rising-sinking motion of the water makes bottom waters inhabitable by tullibee.

I took the last of the wriggling silver-sided fish from my last net for the year and held him, quivering in my hand. What a miraculous sense of relief the lake's overturn must have brought him. I looked into his glassy black eye and congratulated him on surviving the great tribulation that killed so many of his kin. Hot summers come and go. Hopefully next year would be easier. He wiggled. I wished him well and slipped him back into the lake.

Nets stowed in washtubs, I turned the boat back toward the boathouse. I returned the next summer to conclude my studies and see my understandings confirmed.

. . .

A year passed, then two, and, thesis written, Lake Itasca and the tullibee's world faded from mind as I established myself teaching and married Geri, whom I met my second year at the station. Decades passed before Geri and I returned to the station on the occasion of the ninetieth birthday of its establishment.

Enveloped in nostalgia, we walked the familiar grounds once again, past the student cabins, the mess hall, and the flagpole where we first met. We eventually found ourselves at the dock, the launch point for my tullibee studies and memorable moonlight paddles together. Hand in hand we looked out into the water. Something was wrong. This water was murky. This was not the water I knew when plying the lake with my nets and sampling gear those years ago. I stared in disbelief. What could have happened?

I understood that ongoing nutrient enrichment of lakes resulting from human activities was widespread, and many lakes are less clear, murkier, than they were when I did my studies. But here? Lake Itasca? Protected within the State Park?

Later in the day we went on a pontoon boat excursion with others attending the birthday celebration, including the station's resident biologist. "Jon," I said, "Itasca's waters don't look nearly as clean to me as they did back in my research days. Is my memory playing tricks on me or have things truly changed?"

"No, you're right. It's real. Limnologists have noticed. It's nutrient enrichment, but nobody is exactly sure why."

I was dumbstruck. Lake Itasca sits high in its watershed. Elk Lake drains into Itasca, but its shoreline is completely natural, with no cabins, homes, or farmlands from which extra nutrients might come. Could it be activities at the campground? The beach? The station itself? The ghost of improperly handled sewage from park visitors long ago? The atmosphere?

Regardless of its source, the change bodes ill for tullibee. Murkier, more nutrient rich waters mean more food falling to the lake bottom to feed bacteria. More bacteria—more demand on summer oxygen supply—tougher times for tullibee. Could it be that my netting days, with their temperature and oxygen stresses, are now viewed as the good old days by the descendants of fish I caught?

Reports of deteriorating water clarity through nutrient enrichment, cultural eutrophication in the lexicon of limnologists, come from lake watchers across the country. This increasing nutrient supply seems a universal trend accompanying lake shore development, the accelerating human embrace of the lakes we love. Expanding zones of oxygen-depleted bottom waters seem the inevitable result, pushing tullibee farther up into the warm surface waters.

Something else has changed as well. Reports indicate that populations of these fish in the southern edges of their distribution in Michigan, northern Indiana, Minnesota, and Wisconsin have declined in numbers and even disappeared from some lakes in which they once thrived.

I worry about the future of these fish and other cold-water species. With slow but unrelenting increases in temperature, and nutrient enrichment, these fish will face narrower zones of refuge from asphyxiating waters below and cooker temperatures above than their great-grandparents knew. The underwater world of the tullibee has become conjoined with the world of people and global warming.

At the end of my studies of Itasca tullibee, I understood their story to be one of numbers, of temperature degrees and parts per million. I was flat-out wrong. Though tullibee did not inhabit the lakes around Concord in his day, Thoreau's take on lakes was more insightful than mine. He described the lake as "the *landscape's* most expressive feature." Lakes now appear even more deeply rooted in their landscapes than Thoreau could ever have imagined. The tullibee's world in Lake Itasca is a narrative of events far beyond its waters, an archive of events a world beyond.

So tullibee, those meek, silent victims of changing lakes, face an uncertain future, their slipping fortunes all but unnoticed, a casualty of frayed strands in the web that enmeshes us both. Wherein lies the tullibee's hope?

Natural selection masterfully outfitted them for success in the cold, oxygenated depressions left by glacial ice. Those dead fish I picked up may have lacked the adaptations to survive the intense summer stress, but many others survived to pass their genes for greater temperature tolerance on to the next generation. Over time, natural selection might produce a tolerant population. But adapting a genome to new conditions takes a long time. It becomes a race between the speed of change and the speed with which mutation and selection can sufficiently remodel a genome. If natural selection wins the race, the population is delivered from its troubles and continues on into its changing world. If the speed of change wins the race, the tullibee are done.

The tullibee's hope ultimately rests with us, though I wonder if we will concern ourselves with the fate of creatures we largely cannot see. Perception rules the behavioral roost. I expect it would matter if we could understand them not as objects but instead as part of a system where, when something untoward happens to one element we understand there are reverberations for all. Can the perceptual centers of our minds come readily to such systems thinking? It takes a change in habits of mind.

· · ·

I recently read of experiments in Saskatchewan where aerators were installed in several lakes where tullibee suffered the same stresses experienced by Itasca fish. Researchers discovered that tullibee in oxygenated basins experienced a greater volume of comfortable habitat during the stressful times than in basins without such technological help. As I read the account, two images slipped into my mind's eye. The first, an image of a patient in a hospital with oxygen tubes up his nose. The next, a tullibee wearing a respirator. Ludicrous? Image of the future? I can't decide.

Riding the West Wind

It is easy to long for the clarity of the early days of the modern environmental movement when the problems could be seen and smelled and the villains were obvious.

JAMES SPETH

Voyageurs National Park runs some thirty-nine miles east to west along the International Boundary between Minnesota and Ontario. Over a third of its surface is water. Sandwiched between immense Rainy Lake to the north and Kabetogama Lake to the south, the massive Kabetogama Peninsula forms the interior heart of the park. Dozens of small lakes and ponds dot the thirty-mile length of the peninsula. Ryan Lake lies deep in its pristine heart.

In the crisp, mosquito-free air and blazing color of early October, Geri and I recently paddled the twenty-two miles from the road head to see this paradoxical lake for the first time. The lake is not renowned for exceptional natural beauty, great fishing, historical significance, or other conventional attractions. In fact, when I asked a park employee about its location, she said she wasn't familiar with a lake by that name in the park.

Though perhaps known only by lake researchers, Ryan carries an ig-noble distinction. The pike in this lake contain the highest concentration of mercury of any yet tested in Minnesota. Nothing disquiets the rational corner of the mind like irrational contradiction. Why Ryan? Why in this remote corner of the state? My need to understand became compelling.

Ryan Lake lies a day's paddle, three lakes and two portages, away from the Ash River Landing. A gusty headwind roared out of the west as we rounded the east end of the Kabetogama Peninsula and headed into the big waves of Rainy Lake for the final miles to the portage to Ryan.

We crept along, ducking behind islands and protective points when we could, and paddled hard when we could not. The bow rode up the front of the swells and careened down the back side, as though caught in the spin and pull of a yo-yo string. A sandbar between a small island and the shore sapped the waves of their energy and allowed us a peaceful landing at the portage.

A path began a gradual ascent beside a tiny brook that fell silently from pool to pool, meandering in idyllic innocence among spruce and fir and aspen, past mounds of deep green moss and the yellow leaves of autumn. Unbent grass growing in the path and a pile of wolf droppings packed tightly with hair confirmed the wild, remote character of this far corner of Minnesota's north woods. In less than half a mile we descended a brushy hillside to the tip of the small lake.

An aluminum punt lay upside down a few feet away. Stashed there by park officials as a convenience for fishermen, I assume, the small boat eliminated the need to portage one's own craft to the lake. Sedge clumps grew beside it undisturbed. We reloaded the canoe and paddled to the campsite down the shore.

Ryan resembles many lakes that dominate the landscape of the international boundary west from Lake Superior. A mixed forest of pine and spruce, aspen and maple, covers the slopes leading down to the lake. A bald rock outcrop, clothed in cushions of gray and pea green lichens, plunges to the water's edge on the opposite shore. The water was clear and, as true of many lakes of the north, deeply tinted tea-colored brown. The picturesque lake and its pristine setting gave not the slightest hint that something was amiss.

• • •

I first encountered mercury as a boy. Once, while I was rummaging with playmates through a trash heap behind a grocery store in my small town, searching for glass objects to use as targets for rock throwing contests, the store owner caught us. He seemed less upset by the scattered shards of glass we had created and more concerned that we should *never, ever* break the long fluorescent light bulbs that lay mixed with the rest of the trash. "There's a bad poison inside these tubes, boys," he said, as he held one in the air. "It's dangerous to breathe the vapors inside. Mercury."

The second encounter occurred when a friend brought mercury to school, salvaged from a broken thermometer. Beads of this strange metal rolled around like sluggish BBs in his jar. We rubbed it on coins, turning them bright, slippery silver. I made no connection between the two incidents at the time.

It took a personal story an Ojibwa man shared with me to finally give mercury meaning. Years ago, Bob and a helper were netting tullibee on a reservation lake in the fall when the fish came to the shallows to spawn. Smoked, tullibee provide a tasty addition to the winter diet. The tullibee catch went well. The net also occasionally snared large northern pike. Fried pike is one of Bob's favorite foods. He had all he could eat, even after sharing fish with others. He ate pike for breakfast, lunch, and supper for a week and a half, he says. Then he cut back to one pike meal a day, then one a week to stretch his supply.

Four weeks later he felt tingling, then numbness, in his left hand and arm. It spread to his right side, then into his legs and feet. At first he brushed it off as perhaps a pinched nerve or simple muscle ache. When he began having difficulty speaking, he feared his growing symptoms might be those of a stroke. "I had to think through every action before moving. For me to pick up a cup of water was a very deliberate process, and it was particularly difficult if the ground was uneven, because I had reduced sensation in my feet and legs."

He finally ran out of pike. Several months later the symptoms began to subside, the cause remaining a mystery. Coincidentally, months afterward, the director of the Indigenous Environmental Network contacted him about helping with a mercury education effort to alert indigenous people to the risks of eating mercury-contaminated fish. As he read information on the subject, he discovered a description of mercury poisoning. "As I read through it, I [was] stunned," he said. Many of the symptoms that I had experienced [after eating] all those large fish . . . were symptoms of mercury contamination."

．　．　．

That Ryan's waters produce fish contaminated with mercury would not in itself come as great surprise. Mercury in fish at levels posing risk to human health is widespread across the continent, most notably in the surface waters of Minnesota, Wisconsin, Michigan, and Florida, and Ontario and Quebec in Canada. It is also of grave concern from Louisiana and Mississippi east through Georgia and Alabama to the Carolinas, Kentucky, and Tennessee. In all, warnings about unsafe levels of mercury in fish have been issued for 35 percent of our nation's lakes. Twenty-two states have issued warnings advising people to limit the number of fish meals eaten for selected lakes and certain-sized fish.

Our health problems begin not with atoms of mercury alone, which are comparatively unreactive. But when mercury is joined by a gang of atoms called the methyl group, one carbon and three hydrogens, metabolic hell may erupt. Once methylated, mercury can readily enter cells and is inordinately attracted to DNA. There methylmercury interferes with normal cell division, the copying of chromosomes, and the timely creation of needed proteins. The developing nervous system of an embryo or fetus is most vulnerable to its ravages. Methylmercury moves readily from a mother's blood into her fetus, where it can derail normal brain development, leading to a child born with reduced attention span, impaired memory, impaired coordination, damaged visual spatial skills, and impaired neuron pathways required for mastering language. The National Academy of Science estimates that 60,000 babies are born each year in the United States at risk of developmental damage from mercury.

．　．　．

We looked out on the quiet lake surface and the wild shore beyond from our cookfire at the water's edge. The question hardly needed asking. Where did the mercury come from? One's intuition leaps first to polluting factories. But there are no effluent pipes or smokestacks releasing foul chemicals in the neighborhood.

The chemical character of a lake reflects the nature of the rocks and soils in its watershed. An early, tentative explanation of mercury's source in lakes was weathering bedrock. That explanation cannot be the case for Ryan. Other lakes in the park, located in the same bedrock formation, have much lower mercury levels.

That lake watchers should turn last to the atmosphere as a possible source of mercury, after other hypotheses came up dry, is not surprising. The idea seems far-fetched, despite that mercury has long been known to readily evaporate. But mercury, the heavy runny metal of common experience, wafting about in the air like gossamer spider silk? Transported great distances? Ridiculous.

Invisible to the senses, such truth defies conventional perception. We now understand that lakes and their watersheds receive over 90 percent of their mercury from the atmosphere. Humans release four times as much mercury into the air as does nature by incinerating sewage sludge and household garbage, smelting scrap metal, burning trash, and improperly disposing of fluorescent bulbs. Burning coal releases by far the most.

But just when explanations begin falling into place, consider this: Cruiser Lake, another small lake in the Kabetogama Peninsula, lies a mere four miles southwest of Ryan as the heron flies. Water samples reveal that the total mercury concentration in Ryan Lake's summer surface waters exceeds that of Cruiser by six and a half times, and Ryan's concentration of methylmercury produces fish that trigger a health consumption advisory while Cruiser's methylmercury concentration falls below the detection limit.

In all, water samples from twenty small lakes in the park have been analyzed, and the results are equally perplexing. Brown, O'Leary, Little Trout, and Mukooda are like Cruiser, methylmercury below the detectable level. Quarter Line and Shoepack, like Ryan, have methylmercury levels almost ten times as great. Quill, Ek, Locator, and the rest fall somewhere in between. How can lakes so close to each other be so different? What about Ryan makes it stand out? That's why I had to see the lake for myself.

"What do you expect to see?" a skeptic might ask. A fair question. I understood the cycles through which mercury travels in the biosphere, insofar as its complexities were known. I understood that during its travels mercury atoms interact with airborne particles and rainstorms and living things—mercury all quite invisible except to scientists with sophisticated analytical tools. What did I expect to *see* indeed? Perhaps curiosity and our unshakeable faith in vision need no further reason.

· · ·

As Geri organized a kitchen around the fire ring, I got my fishing rod and a red-and-white daredevil. I cast from a low rock and watched for the lure's flash of white as I reeled it in. Ryan's water is dark maroon in color. I retrieved my lure almost to my feet before I could see it. I began wondering if tea-colored water might serve as a fishermen's warning of a lake's potential for having high levels of mercury in its fish.

There is a logical connection. As a gardener and backyard composter, I know dead plants rot to become piles of rich, dark compost. A similar process occurs in wet environments. Large organic molecules called humates compose most of the organic material in those rich dark piles. The black *humus* of potting soil and forest floors is well named. Humates readily adsorb molecules to their large surfaces, and activated mercury atoms are especially attracted to them. Water draining into a lake carries humates, bringing mercury along. The darker the water, conceivably, the greater the mercury content.

The average visible depth for the five lakes with the lowest methylmercury content was three times greater than for the five lakes with the most mercury. Hypothesis supported. Then, I noted, lakes with mercury below detection limits were also over four times *as deep* as the others. High mercury levels correlated with more than brownness of the water. Hypothesis dashed.

· · ·

I raised my fishing rod and cast again. The line pulled taut, then went limp. Two casts later, another strike. The hook held and I reeled in a two-pound northern pike. The lure's flashing patches of white and red proved irresistible for the pike. I grasped the fish tightly and lifted it from the water, daredevil dangling from its lip. Attack first and deal with consequences later seems the credo of this species.

How delectable golden brown fillets fresh from the lake would taste. My torpedo-shaped fish's shiny black back, white underbelly, and whitish splotches along its darkish green sides looked like those I'd caught for years. But looks deceive. The fish consumption advisory for Ryan Lake says I may safely eat only one meal of such creatures in a month's time. If my wife were pregnant or we had a small child along, they could not safely eat it at all.

An insidious agent, mercury operates under the radar screen of our perceptions. Eat a plateful of contaminated fish and nothing happens. You don't die. You don't even feel sick.

Geri and I could choose to eat pasta for supper and return this fish and its mercury to the lake with little consequence for our tummies. But the reality is different for native peoples scattered from the upper Midwest to the arctic coast, steeped in a culture of subsistence hunting and fishing. Years earlier, Canadian health officials became aware of unsafe mercury levels in Indian peoples in northern Ontario. What does one say to a community that eats fish not once a month but many times a week, and in doing so exposes their children, even before birth, to the neurological ravages of mercury tagged with the unholy methyl group? Do you advise such people to stop eating the fish and so destroy a lifestyle, disrupt a culture, and wreak economic and social havoc? Do you refrain from giving such advice and creating such disruption and let the children grow to become less than they could be? For the communities of Grassy Narrows and White Dog in northern Ontario, the choice was disrupt the culture, protect the children.

The fish wiggled against my grip, betrayed by flashing metal he did not understand. For as long as people have lived beside and traveled lakes, we have trusted the waters and eaten the fish. Now the trust lies broken, betrayed by events we did not understand a thousand miles away. I unhooked the pike and released him back into the lake. His black back disappeared instantly in the dark water.

Though elemental mercury falls on Ryan Lake from the skies, the studies are clear: absent connivance with the methyl gang, it would not put fish eaters at risk. Lakes like Ryan and their watersheds apparently facilitate the devilish match-up of methyl to mercury more effectively than other lakes. But what causes the difference?

A critical clue came from Florida. Uneasy about reports of fish bearing inordinately high levels of mercury in the upper Midwest, Canada, and northern Europe, Florida scientists began testing their own fish and discovered most specimens registered mercury levels just barely within acceptable limits—until they went into the Everglades. There they discovered fish contaminated at levels twice that elsewhere in the state, and as high as the most contaminated fish yet discovered in the nation's freshwaters.

The highest concentrations of humic substances are found in bog waters, marshes, and wetlands. Plants here have perpetually wet feet and create conditions intimately linked to the mercury story.

Other studies show that wet places provide precisely the conditions required by a particular breed of sulfur bacteria, bacteria that, in the course of their body metabolism, attach methyl groups to mercury atoms forming the nasty methylmercury. Based on these discoveries I came to Ryan expecting to find it surrounded by extensive wetlands.

Ryan is roughly a half mile long and a fourth that wide. The next morning we set out to explore the lake's shore by canoe. Insignificant stands of grasslike sedges, a few small clusters of watershield plants, and tiny slivers of bog mat grew in scattered places along the mostly bedrock shore—little habitat for methylating bacteria. The lake tapers at its southern tip, and its basin continues south as a long narrow bog. I left Geri with the canoe and walked the bog's spongy surface, water seeping into my footprints as I stepped. Bacteria were creating methylated mercury directly beneath my feet.

Though the bog accounted for a small portion of Ryan Lake's shoreline, it receives drainage from a larger area watershed, increasing its methylmercury intake. Beds of deep green sphagnum moss fill spaces between thinly spaced conifers on the steep west shore. Moss also provides the wet acidic conditions so to the liking of methylating bacteria. Rain flushes that methylmercury down the slopes to the lake, adding to its methylmercury load.

· · ·

Eating pike from Ryan would be a particularly stupid thing for a pregnant woman or a child to do, but, paradoxically, either could drink quarts of the lake's water quite free from harm. The bog water that oozed beneath my feet contained a vanishingly small concentration of methylmercury, a third of a billionth of the weight of a paper clip in a little over a quart of water. Though for lake water this is high, the flesh of the fish I caught likely had a concentration three million times as much.

It can be explained. Methylmercury slips readily into algal cells. Daphnids and other animal plankton eat algae and retain the mercury, accumulating it into higher concentration. Small fish eat daphnids further concentrating the mercury. Many of these molecules eventually reach the flesh of the top predators in the lake, including northern pike. The pike's vicious lunge, performed successfully a thousand times before on more palatable prey than my daredevil, accumulated ever higher mercury concentrations with each bite, enough methylmercury to discombobulate new cells in a developing human brain.

. . .

Directly across from our campsite, a tiny bog filled an indentation in the lake's shoreline. We investigated. At the water's edge and midway down the bog's length, the random pattern of sedge stems was suspiciously interrupted. There, partly hidden by a thin sedge curtain, plant stems lay flattened, forming a roundish pad several feet across, the past summer's nest of a loon. Fish eaters both, loons and people sit at the top of the lake's food chain, the final destination of methylmercury in its sojourn in the lake. The basic body chemistry of loons hardly differs from that of people, and methylmercury is potentially as devastating for baby loons as it is for baby people. Sadly for loons, many of them choose to nest on northern lakes whose fish contain bioaccumulated mercury. We saw no loons on Ryan Lake during our two days there. October is late in the season. The pair that had nested on the sedge pad had likely flocked-up with others and migrated south for the winter. Had any young birds accompanied their parents this year? The odds were not good. Loon pairs nesting in places with the mercury content of Ryan often make the fall flight alone. While we chose not to eat the pike I caught, loons could not understand the danger.

. . .

Back at camp, fecal material jam-packed with indigestible crayfish parts lay on the rock a foot from the water's edge, the droppings of an otter, another creature high on the food chain. The discovery some time ago of a dead Florida panther, killed by mercury poisoning, revealed that even predators who don't eat fish, but eat other foods high in the aquatic food chain, are vulnerable. Raccoons had served as the mercury conduit from the food web of the swamp to the tissues of the panther. The higher the mercury levels in their lactating mothers, the lower the survival rate of panther young. Our otter probably picked up a hefty mercury dose from all those crayfish bodies he had digested.

We broke camp and retraced the portage trail to Rainy Lake. The west wind blew and waves pounded even more vigorously than when we had come. The sandbar that had protected our landing going into Ryan protected our launch as well. But the gusto of the big lake was overwhelming. Using the sandbar as a shield from the largest of the waves, we paddled the hundred yards to the tip of the small island, then pulled ashore to wait out the blow. The wind's continuing presence became an emphatic reminder of its other link to our lives.

Whitecaps charged out of the open horizon of the Brule Narrows nearly twenty miles away and raced down the lake as far as the eye could see. I sat on a rock, back from the spray zone, and watched. Weather systems move from west to east in this part of the country. The western sky is the gateway from which mercury often comes.

The day began clear, but by afternoon clouds appeared and rain squalls began zipping east across the sky. One raced the length of the lake, dumping its mercury scrubbings onto Canadian waters. Another, with abject disregard for the international boundary, stormed out of Canada making landfall two miles down our shore, dumping its mercury in a path of invisible confetti onto the lake and the wetlands of the Kabetogama Peninsula.

Where did today's wind pick up its mercury? Seattle and Tokyo by way of Winnipeg? Does it even matter? Mercury atoms can ride the winds for a year before landing. Sources? West becomes east. North becomes south. Another squall made a beeline for our island and Ryan Lake. I hurried to get my raincoat.

How strange, how improbable it all sounds. My mind pleads for the visible, the concrete. I must drag it from its reluctance to comprehend such complicated truth. Our species came of age never having confronted poison from half a world away falling from the sky. Might our understanding of the connectedness of distant places be more intuitive had we the perspective of the migrating loon, carried on its highway of air? Though earthbound no more, our minds still tilt to the winds of our direct senses. We must see and taste and smell.

I left the gusty wind on the island's west side mid-afternoon for respite on the lee shore. There, propped up on my elbows, I watched quiet water and noticed an otter thirty yards down the beach coming my way. Her sleek brown body cavorted smoothly, flowing beneath the surface, then up, then under. So effortlessly she swam, as though reveling in the pure joy of movement. At times she dove, surfaced, and floated on her back, manipulating a clam or crayfish on her tummy.

She got closer. How magnificent she was. As she closed the distance between us she moved a bit farther from shore, perhaps wary of my presence. She continued past me, then struck out across the bay and disappeared into the distance, likely heading to other hunting grounds. Otters love to eat fish. I could not help but wonder, as she disappeared across the bay, about her babies.

One thing we humans don't share with loons or otters: we can read fish consumption advisories. That matters only if people know of these warnings and their significance, a circumstance less likely for poor, less-educated people.

The wind lost none of its ferocity that day, and we camped overnight on the island in a sheltered pine grove. A lull in the wind at dawn allowed us to leave for home.

. . .

In a Florida study, 405 people were identified who had eaten fish or wildlife from the Everglades at least once a month, putting them at risk for mercury damage to their bodies. Because of their risk, investigators asked these persons to participate in further studies to help them identify more precisely those subpopulations at elevated risk. Fifty-five of them flatly refused. When asked why, the most common reply was that such participation would interfere with their fishing.

Ah, the human mind. If we can't smell it or see it, can we trust that it even exists? Such is the fate of things blowing in the wind.

FUTURESCAPES

Shield Lakes Icon

The spectacular fish is a memory of its past and a vision for a desired future, an icon to stir human action on behalf of valued and relatively unspoiled Boreal lakes.

JOHN J. MAGNUSON

North of a line from Maine through the Adirondacks west to Minnesota then northwest to the Beaufort Sea, the great ice sheet scraped and gouged the earth's surface like an unrelenting bulldozer. In doing so the ice laid bare the ancient rock core of the continent, the Precambrian Shield, and gave birth to a vast number of lakes.

In the course of several million years of ebbing and flowing across the land, the ice also midwifed the birth of a remarkable fish, *Salvelinus naymacush,* the lake trout. For two or three million years, the forebears of today's lake trout moved like summer tourists. They came north, following receding ice when the glaciers waned, then retreated south, east, and west when the ice readvanced. This fish not only inhabits the cold bedrock lakes today, it is essentially confined to them. The greatest density of lake trout lakes on the continent lies along the southern edge of the exposed shield along the Minnesota-Ontario boundary. I learned about lake trout through my studies of tullibee. Both are cold-water fish and often inhabit the same lakes. Unfortunately, my knowledge of these fish is limited to book-learning, gleaned from reading about them in fisheries journals. To understand these bedrock lakes, one must understand the trout.

Geri and I have come to Ontario's Quetico Provincial Park to fill the hole in my understanding through the ground-truthing of direct personal experience. I want to understand these fish and to experience them the way humans have done on these waters for millennia—as prey.

Since I never before fished for lake trout, I sought the advice of an expert. Chuck has caught many lakers in a lifetime of fishing. His advice was simple: use a stiff rod with deep-running plugs, and jigs. "Remember," he said, "It's May. This time of year they can be anywhere in the water column."

We launch onto quiet water under a cloudy sky at the public access on Beaverhouse Lake in Ontario's Quetico Park. This park and the famed Boundary Waters Canoe Area Wilderness on the American side together protect 2,000 lakes and 2 million acres of lake-studded shield in wilderness status. We plan to travel the lakes in the western end of Quetico Park for nine days.

This is an unusual trip for me. Though I always bring fishing tackle along on lake trips, I typically have greater interest in exploration and photography. I usually spend little time fishing and often don't catch anything at all. This time is to be different. I approach the fishing with as much pessimism as hope. Whether due to lack of skill, lack of patience, or plain bad luck I cannot say, but my past fishing success has been spotty. I have grown accustomed to failure, to even expect it. My hope this time lies in my commitment to persist, to fish constantly, tirelessly. Maybe by the end of our trip I will have something to show for it.

. . .

We cross Beaverhouse to the Canadian Ranger station for our entry permit and discover the station has not yet opened for the year. Fortunately, a maintenance worker is here tidying things up for the upcoming season. He graciously rummages through desk drawers and eventually finds the proper forms. We return to the water and head east, where a short portage will put us on Quetico Lake.

Once on Quetico I waste no time. I reach for my fishing rod and attach a deep-running lure and pay out yards of line. Rod propped against the thwart with my foot, we begin trolling canoeist-style down the lake. The rod tip vibrates rapidly. All is in order as arms and shoulders fall into easy rhythm. A breeze pushes us gently along.

I now notice the rod's tip has stopped vibrating and its slight bend has disappeared. I reel in a lureless line. I had felt no snag. Had a northern pike taken a swipe and overbit, severing the line? Had a failed knot let a trout escape? Who knows.

I switch tactics. Letting the breeze move the canoe into a bay, I begin jigging. In twenty minutes I have a fish on the line. The closer I play him to the canoe, the harder he fights, swerving, diving, his black back breaking the water surface.

"That's a nice fish," Geri exclaims. "What is it?"

"Not sure," I reply. The fish surfaces again.

"Looks like a northern," says Geri.

"Nope," I shoot back. "I saw an adipose fin. It's a trout." He takes several final desperation dives before I maneuver him to the side of the canoe. His gills pump in exhaustion. I reach down, grab the fish firmly in my hand, and lift it over the gunwale. At that moment the jig pops loose from the fish's lip.

I stare at the fish lying at my feet, stunned. How beautiful his vibrant pulsing body, scales and white spots on his side glistening in the sun. This wasn't the way it was supposed to happen, barely hours into our trip. This was to be a long drawn-out quest. I am both shocked and elated with such immediate success. "Fantastic! I got a trout," I shout. Geri chuckles at her babbling partner.

Suddenly, the inexplicable happens. A strange hesitancy, uncertainty, floods my mind. I look down at the fish. Excitement ebbs and jubilation turns into remorse. I don't want to see him die.

Me, of all people? Sad about taking a fish? I have successfully hunted deer and moose and caught most of the common game fish. The fish flops, slapping himself against the bottom of the canoe. I quickly, instinctively, press a hand against his quivering body to prevent him from launching himself overboard, and look at the creature transfixed.

Time passes. I slowly pull the stringer from my tackle box, push its metal lead through the fish's mouth, tie the end of its cord to my seat, and slip the fish into the water. We paddle slowly toward a spot on shore that looks suitable for a camp.

I don't understand my strange reaction to this fish. Should we let him go? Should we keep him? I sense Geri already has visions of trout for supper.

We land and unload packs but leave the canoe floating in the water, fish attached. After we erect the tent and put camp in order, Geri pulls out her book and begins reading. I walk to a log, sit, and stare at the canoe, my mind refusing to leave the image of the fish on a leash.

I step from my log and wade into the water. The trout rests quietly, hunkered beneath the boat, gill covers pumping in slow rhythm, his shadow silhouetted against the sand colored bottom. He is a solid, stocky fish, thick across the back.

I turn, take anguished steps up the beach to Geri, and break the silence. "That fish is too big for us to eat for supper. No way can I justify killing him and throwing half away. We should let him go."

Geri silently acquiesces. I return to the canoe stern, untie the stringer, carefully withdraw it from his mouth, and give him a push. He swims strongly from the canoe and disappears. Uninjured by the hook, he will survive. My remorse turns instantly into relief and I return to my log.

What explains my reaction? I've netted tullibee by the dozens. Was the prize too easily gained? Most hunting and fishing friends of mine feel remorse at taking the life of a wild creature. I know that feeling, but it runs more deeply now. Is my reaction actually disguised disappointment at besting an opponent who turned out to be much less formidable than expected? I doubt it. I fish to eat, not to conquer. A self-searching hour later, I decide to try to catch something else for supper. I return to the canoe and paddle across our small bay on water glass smooth.

. . .

One cast and bam! A fish explodes the surface, engulfing my plug the split second it hits the water. I haul a northern pike into the canoe, the perfect size for supper.

As I paddle toward camp I notice a cow moose standing in the water far down the shore, watching me. I paddle slowly, quietly toward her. She does not spook. I reach her. I'm as close as prudence will allow. She looks calmly at me and remains motionless. The canoe drifts past her. She ignores me as if I were a lily pad. Now a pair of loons pop up beside me, white-and-black bodies and red eyes glistening in the evening glow of waning sunlight. They sit motionless and watch me. I am enveloped in magic.

According to Richard K. Nelson, writer and anthropologist, the Koyukon peoples of Alaska believe if you treat animals with respect they will come to you, offer themselves to you, the hunter, the fisherman—but only if the spirits of hunter and prey are properly aligned. The string of magical moments with pike and moose and loons so soon after releasing the trout feels like such a mystical sign. It is as though by freeing the trout I have allowed the collective spirit of the animal life of this place to join with my own. They are giving themselves to me.

The loons now dive and the moose ambles back toward the woods. I turn the canoe toward camp, aglow. I have not the slightest misgivings about cleaning my pike for supper. I cannot explain why pike is okay but not lake trout. The pike tastes superb.

I sleep little, puzzling. I admit to feeling deep admiration for any living creature that succeeds in a challenging environment. Beautiful and inviting as the shield lakes appear to a paddler, conditions are austere beneath the surface. Cold, phosphorus-poor and unproductive, food webs skimpy and simple, these granite swimming pools are an unyielding place for a fish to make a living. Maybe my admiration for the trout's adaptations to these stern conditions has drawn out my sympathetic reaction.

Lake trout have a modified metabolism that enables them to grow more efficiently in cold water than other species. They are opportunistic eaters. Although they do best on a diet of fish, they will eat almost any available animal food: freshwater sponges, crustacea, aquatic insects, and land insects that happen to fall into the water, and even plankton if need be. They are large, long-lived fish that can survive long periods without food. They also provide their eggs with large quantities of nutrients so offspring get a boost as they start life in these inhospitable waters.

Though the trout is beautifully adapted to the cold sterile lakes, maybe its vulnerabilities made it so hard for me to kill my fish. The threats to lake trout are mostly of our making and nearly all of them unintended.

Take the competitive smallmouth bass, for example. Smallmouths are not native to these shield lakes. For decades well-meaning fishermen carried fingerlings of these snappy fighters across watershed divides north into new waters, into lake trout country. Boon to bass fishermen, the intruders are bane to lake trout fishermen. Diets of bass and trout overlap. Trout can typically hold their own where populations of tullibee, whitefish, and smelt are strong. But in other, generally smaller lakes, where trout rely more on near-shore fish prey, smallmouth bass can disrupt their food supply. Trout could, in theory, shift to eating plankton or other nonfish food under such conditions. But fish forced into such a diet grow more slowly, remain smaller, and die younger than those able to feed on fish. Official stocking of smallmouths has stopped, but the species continues to expand its range into more trout lakes.

.   .   .

Several days later I equivocate about letting the fish go. Was it a foolish thing to have done? Might I catch another one, smaller, that we could eat with no waste? Ancient instincts rooted in an ancestry of hunter-gatherers rise within me. I give in and resolve to resume fishing for trout.

We portage out of Quetico into another long skinny lake. Though maps show these lakes to be irregular in shape, many are long and narrow, with most running nearly parallel to each other in a northeast-southwest orientation. Long parallel scratch marks on rock outcrops run in the same direction, the glacier's grinding path written indelibly on the land and map alike.

Shield rocks are extremely old, highly resistant to weathering, and contain miniscule amounts of phosphorus. Sparse phosphorus means sparse plant growth and sparse algae, fewer waterfleas, a bare-bones food chain, and clear water.

We paddle along shore. A lake bottom covered with large angular rocks glides by, ideal spawning grounds for lake trout. I resume fishing. Half an hour later I catch a trout, smaller than the first but enough for supper.

Ontario fishing regulations limit take to one lake trout per day. I understand why. These fish are much more vulnerable to overfishing than other species. They grow more slowly, taking much longer to reach sexual maturity, and so take longer to replace themselves than other fishes.

A gull discovers us as soon as I begin gutting the fish. He lands on a rock and watches. The fish's flesh is white. Lake trout flesh can be any of a variety of colors, including orange and even pink. Her stomach contains eight partly digested gold and brown terrestrial leaf bugs, identical to live bugs I saw struggling on the lake surface just down the shore. The anxious gull makes short work of the gut pile, enjoying his trout dinner, I expect, as much as Geri and I.

I suspect my good fortune catching trout has less to do with my fishing skills and more because trout are distributed throughout the water column. I might not be so lucky in midsummer, when surface waters warm. Like tullibee, lake trout are happiest in cold water, with a lake temperature of fifty degrees or cooler most to their liking. Like tullibee, trout descend to deeper, more comfortably cool water as the summer progresses and surface water warms.

Unlike Lake Itasca and the large majority of lakes to the south, the nutrient-poor shield lakes with sparse plant and animal life provide little food for bottom-dwelling decomposition bacteria. Few bacteria mean oxygen concentrations remain high enough in bottom waters to satisfy fish all summer long. Lake trout, preferring dissolved oxygen at no less than six parts per million, would never survive the summer hell tullibee regularly face in the bottom waters of Lake Itasca.

A new day begins under clouds that look like they want to rain. About noon, in a fine mist, I land another trout. Sadly, over half of western Ontario's trout lakes have a fish consumption advisory for mercury. Is it safe for us to eat another? Because of bioaccumulation, the larger a fish, the higher its mercury content. Though of good eating size, my fish is certainly no monster. Since we will not be eating trout several times a day for weeks on end, we decide it is probably safe to eat this one.

Mist turns into light rain and light rain into hard rain. The wind increases and we begin looking for a campsite. The first spot we come on is already occupied. We continue searching and find a suitable rocky point, pull into its lee cove, and begin hauling packs to an open stand of red pine in steady rain. Two herring gulls land on a rock thirty yards from the canoe and watch. Not until I trudge off with a pack and the gulls make a beeline for the canoe's stern do I realize they are more interested in my fish hiding tethered beneath the canoe than in water-logged canoeists.

Hard rain continues. We grow hungry for supper and for trout. Unfortunately, I did not bring our camp stove. That was deliberate. Given the long portages on our planned route, I had shed as much weight from the packs as I could. Besides, I take pride in my ability to start a wood fire under even the wettest conditions. But faulty rain gear has me holed up in cold wet clothes in the tent.

As heavy rain dances off the tent fly, I think about my trout hiding helplessly beneath the canoe. What exasperation he must feel, leashed like a dog to a piece of Kevlar from a world not of his making. My thoughts turn to other intrusions into his world. Global warming can bring no good to a cold-water fish. Research shows that adding global warming to the mix of ozone depletion and acidic rain results in a tragic synergy. Methylmercury increases and lake trout habitat gets squeezed. Some researchers expect that under current climate change models, the smaller, more shallow trout lakes stand to lose a substantial proportion of existing optimal habitat.

With no fire we cannot eat the fish for supper. We could eat him for breakfast. Then I think of the gulls waiting for him to make a false move. He is probably safe from the birds shielded by the canoe. But what about otter or mink or even a large snapping turtle? They'd relish that trout as much as would gulls. Constrained by the stringer, my fish would not have a chance. What an ignoble end for the icon of shield lakes. This fish and his kind face problems enough. He doesn't need one more from me.

In bare feet, soaked underwear shorts, and a raincoat, I unzip the tent and pick my way through the rain to the canoe. The fish is feisty as I wade into the water, undo his leash, and aim him toward open lake. A powerful swish of his tail and he is free.

I run through pelting rain back to the tent, remove my soaking clothes and shove them out through the door flap. I'll deal with them tomorrow. Dry clothes on, we break out supper—rye crisp, cheese, gorp, and energy bars—and eat to the sound of roaring rain.

"Feel better about the fish?" Geri asks.

"Yes, he'll be okay," I reply. I put on sleeping socks and an extra shirt against the chill and crawl into a dry sleeping bag. How comforting the refuge . . . Refuge . . . My just-released fish now wanders in his own refuge in the cold water of the lake.

Years ago I worked with many others to establish a refuge, a sanctuary in the form of official wilderness designation for lakes and lands and such creatures as trout, on the American side of the Quetico-Superior lakeland. When Congress passed the act, we saw it as establishing refuge in perpetuity, unless undone by those who make laws. But thirty years ago I never dreamed of mercury and global warming.

I think about trout and bedrock lakes of the northern parts of Wisconsin and Michigan and southern Ontario and Quebec that lack the protections present in the Adirondacks and the designated wilderness refuges of Ontario and Minnesota. Shield lakes, with their clear water and rocky shores, powerfully attract humans and development: roads, houses, lawns, driveways, phosphorus runoff, lowered water quality, removal of shoreland habitat, and warming water. So less certain are their futures.

The world of the trout is becoming much different from the one in which their tribe arose and to which they are so uniquely adapted. While we do not yet fully understand all the impending consequences of global warming, deeper ultraviolet penetration into the water, acidification, new species introductions, and mercury, one thing is certain. The shield lakes and this extraordinary fish intimately share a past, a present, and a future. The future of both shield lakes and lake trout will be one and the same.

• • •

Back home in the contemplative quiet of my writing shack I have arrived at a tentative explanation of my puzzling response to that first trout. While the scientist in me vigilantly guards against allowing emotion to deflect my perceptions of truth, I think in the case of the trout the opposite happened: real world threats to the fish and its home steered my mind to emotional truth.

The Future in a Raindrop

*All we do leaves a mark by which others, now and later, may
understand and judge us. In this respect our art and our lakes
are alternative mirrors to our society.*

BRIAN MOSS

Seven and a half miles of paddling and a mile-long portage, uphill, puts you
and your canoe on Grace Lake. A twenty minute hike up the lake's steeply
sloped north side, longer if you like blueberries, puts you on top of a ridge
of distinctive white rock. From there you can look down on the lake and its
tiny islands and see that it lies in a narrow trough between the ridge you're
on and a parallel ridge on the other side. Both ridges are of the same white
rock—quartzite, geologists call it.

Melt white grains of sand and you get clear glass. Squeeze and cook
white-grained sandstone and you get quartzite. Jammed hard by another
rock mass, quartzite buckles, here into giant waves of opaque glass. Waves
300 feet from crest to trough, and a third of a mile across from crest to crest.
You can see these things now that weathering, erosion, and moving ice have
scraped and worn away the younger, softer rocks that once covered them.

I stand on the crest of one of these milky-white waves. Geri has remained in camp at the lake's shore. I look east down lake through a hazy
sky and notice a slight dip in the horizon. My map shows a portage trail,

more than a mile long, lying in that slot. That path leads to Nellie Lake, our destination.

The map labels this rugged terrain of southern Ontario the La Cloche Mountains, most of which lie in Killarney Provincial Park. A place called Widjawa Lodge serves as the sole park permit station here in the park's far west end. As Geri and I picked up our permit, I asked Derek, the proprietor of that small mom-and-pop operation, if many people visit Nellie Lake. "Oh, Nellie and Grace are the most popular lakes in this end of the park," he replied.

"What's the attraction?"

"The water. Nellie's water's just unbelievably clear. Nothing living in it. People love to swim there."

I have heard of the incredible clarity of the lake, clearest of a hundred and fifty lakes in the park and among the clearest lakes on the continent. I also know the high price it paid for its status.

The La Cloche Mountains and their lakes made headlines around the world more than thirty years ago. Fish populations dying. Lakes becoming as acidic as tomato juice. Scientists discovered decades ago that sulfur emitted from smelters and coal-fired power plants makes acid rain, and acid rain makes acidic lakes. Researchers soon discovered that bedrock lakes and rivers from Nova Scotia across Canada and New England to northern Minnesota were at risk. The public in Canada and the United States cried out: *Save the fish! Save the lakes!* Sulfur emission reduction laws were eventually adopted in both countries and the subject disappeared from the headlines.

We came to the La Cloche lakes to see how the passage of time has treated this infamous place. More important, I came for a glimpse of the future.

· · ·

A map of the upper Great Lakes shows the La Cloche Mountains lying on the north edge of Georgian Bay at the north end of Lake Huron. Though Westerners might snicker at calling elevations no more than 1,600 feet above sea level *mountains,* these white ridges are the roots of ancient peaks that once towered higher than the Rockies. That there is anything at all left of these mountains after eons of weathering and erosion is a tribute to the toughness of the quartzite. Unfortunately, the same mineral composition making quartzite so resistant to weathering makes it a poor source of molecules able to dissolve in rain water, trickle into a lake, and neutralize its acidic waters.

· · ·

We break camp in the morning and paddle the short distance to the end of Grace Lake. The trail from Grace to Nellie is long and uphill. I lean the bow of the canoe into trees periodically to rest aching shoulders. Geri carries a pack. The path ends on tiny Carmichael Lake. We paddle through a narrow stretch of water and then onto Nellie itself. Like Grace, Nellie is long and narrow and lies hemmed in by quartzite ridges. The lake is stunning in its beauty.

As the quartzite stood by, raindrops delivered their invisible acid to the lake water for decades and so delivered a death blow to its life forms. By 1970 lake acidity had increased one hundredfold. Its fishes died. Its plankton died. Its insects died, and so did its algae and plant life. If to you a lake is simply a body of water, clean and pure, you would love Nellie. But if a lake means lily pads and minnows and gulls and myriad other creatures great and small dancing the grand dance of life, you will have a hard time holding back the tears.

I have never seen water so clear. I have fifty feet of measured line by which to submerge my Secchi disc. The lake surface is calm and the sun penetrates the haze. I lower the disc. Down it sinks, deeper and deeper, the intensity of the white light reflecting off its surface not diminishing one whit as it descends. Rope plays out in loose coils. Five feet. Ten feet. Now fifteen. Suddenly I realize that in my haste to get a reading I have neglected to tie the end of the rope to a thwart. I halt the plate's descent and make the tie, relieved the rope hasn't slipped from my hand, consigning my disc to eternity at the lake bottom. The rope and plate resume their descent. Twenty, thirty, forty, now fifty feet down. The rope ends. Still the white quadrants of the plate shine up at me bright as buttons. "Geri! Fifty feet! And I'm not even close to disc disappearance." Fifty feet is twice the depth of visibility I might find in any other lake I know. Others record the Secchi depth here at over one hundred feet.

A lake usually takes on the reflected color of the sky. Blue sky, blue lake. Gray sky, gray lake. But that is not true here. When thin low clouds pass overhead, shadowing the lake, it remains this striking blue. The color is neither intense nor pastel. It's bluebird blue with the slightest hint of gray tinged with cobalt, an ethereal, otherworldly color. Swimming here must be like floating on air.

Some have proclaimed lakes like this to be dead. But surely no lake can be literally lifeless. Even the incredibly hot pools of Yellowstone, hot enough to slow cook animal flesh, have their flora of heat-loving bacteria. What kind of life *does* live in this "dead" lake?

Ah, a few dragonflies cruise near shore. Small, black bodied, the first few abdominal segments a deep dark red. This species is unfamiliar to me. The six-legged google-eyed dragonfly young spend their lives underwater, but they are not especially tolerant of highly acidic water. Have they matured elsewhere and flown here to feed? Adult dragonflies eat insects. Aquatic insects must be few if there are any here at all. Perhaps they feed on land insects that fly out over the water.

A small swarm of beetles zips about the water surface. By their crazy, erratic motion I recognize them as whirligigs. Whirligigs often live in bog pools, among the most naturally acidic aquatic habitats of all. They would come preadapted to the acidic conditions of this lake.

We paddle the north shore and into a narrow bay running inland from the main lake. Several logs float, mostly submerged, in the bay's shallow water. A crowd of diminutive pin-cushion-like red sundew plants, bristling with tiny stalks, each topped with a glistening drop of viscous glue, form a bright red cap the length of each log. Sundews, like the more famous Venus flytrap, have turned the normal relationship between plant and animal upside down. These plants catch *insects* in those sticky drops of glue and digest them. Considering they commonly grow in acidic, boggy places, I am not surprised to find sundews here. Bogs are often nutrient-poor places where insect-eating habits provide such plants a nutrient supplement. But where are the insects? Obviously the robust little plants have something to teach me about realities in Nellie Lake.

A dragonfly struggles spread-eagled on top of one of the red mounds, an insect Gulliver held fast by a dozen tiny glue-tipped arms. He flutters his wings in panic and lurches forward out of the grasp of some of the arms into the grasp of others. A final Herculean convulsion of wings and he breaks free, barely escaping being digested alive.

We glide deeper into the bay. A shallow bottom of gray silt rises toward the canoe. Protected from waves on the main lake, this bay in countless other lakes would be rife with pondweeds or water lilies or reeds or any number of other emergent or floating leaf plants. There are none. Unfamiliar white jackstraw-like stems lie scattered randomly among small stubby green plants that grow upright from vigorous white roots. The rosette arrangement of their leaves reminds me of the quillworts I've seen in other northern lakes. These plants thrive in lakes too sterile for other plants. They also, apparently, tolerate a degree of acidified water.

Bays like this normally act as nurseries for young fish, but neither waterfleas nor minnows appear in the clear water. No herons stalk, no puddleducks swim, no sandpipers, not even the raucous call of the kingfisher disturbs the lonely, eerie silence.

Blocky chunks of quartzite the size of bricks pave the lake bottom farther down the shore. Some shine brilliantly white through the clear water. But many are the color of rust. All are sharply angled, none have the rounded edges of stones that have been jostled for eons by moving water.

A wide flat exposure of the quartzite forms a clearing at the water's edge, and we pull the canoe onto shore. I sit on the rock and dangle my legs into the diamond-clear water. Inexplicably, under water they turn ghostly white. After lunch I turn submerged rocks over one by one. The underside of each reveals nothing except streaks of thin, collapsed, mucous tubes. Creature activity to be sure, but the menagerie of mayflies, caddisflies, snails, tiny leeches, and crayfish typical of lakes is missing.

The rust on the rocks forms a thin covering that my fingernails cannot scrape away. It must be a chemical deposit, not the periphyton so ubiquitous in normal lakes. No wonder I find no snails.

A green frog materializes in a small water-filled depression in the bedrock, inches from the water and a short hop from my knee. He is two inches long, black spots on brownish back, and tiger stripes on folded legs.

A deerfly begins to make lunch of my shoulder. I instinctively swat the pest and the frog shifts his body to face me more directly. I take the smunched fly by the wing and set it motionless before the frog. No response. Now the fly wiggles. Snap! The frog's tongue, quicker than sight, gulps it down. How long must a frog wait on this shore before lunch comes along on its own?

Since the undersides of rocks have provided such meager information about insect life here, I begin searching shoreline shrubs to see what the spiders, experts on such matters, have to say. I find most plants barren of webs. I now see a large web facing the lake. Two mosquitoes and four tiny midge carcasses hang in the otherwise empty webbing. Many strands of broken silk suggest the owner grows weary of maintaining his unproductive pantry.

Paddling resumed, Geri spots a beaver house, the third since entering Nellie. Though some beaver prefer streams, others choose to live on the shores of lakes. The debarked cuttings dragged up on the house are not fresh but are certainly not older than a year. How wonderful to see a bit of normalcy. The sound of the single pluck on a loose bass fiddle string breaks the silence—a frog up the shore seeks love.

The haze thickens. Might it be smoke from a forest fire? But where's the smell of burning wood? I had hoped to climb one of the quartzite ridges with my camera to capture the stark beauty of this mountain lake on film. Photos under these conditions will not do the picturesque setting justice.

Oh my word! I can't believe it! A loon pair appears some forty yards away. For half a day we have been on this lake and heard not the sound of a single shore or water bird, let alone seen one. I have not been surprised at the absence of loons. Loons eat fish. What in heaven's name are they doing on a fishless lake? They certainly can't have a nest here. You can't raise loon babies on an occasional midge or mosquito or dragonfly. I know loons sometimes leave their nesting lake for others nearby. Is this visit one of reconnaissance to see if conditions have improved since last year? Were they lured here by some tribal memory of delicious lake trout and ciscoes swimming in water so clear they had no place to hide from a hungry loon?

Or have they discovered other food? I've heard that loons will eat crayfish if they can find them, enough to give themselves pink-tinged feathers. Crayfish are more mobile than many other aquatic creatures. Though

I've not seen one, maybe crayfish have returned to this lake, reestablishing populations decimated in the early onslaught of acid.

The loons dive, then surface much closer to the canoe. We drift toward them, yet they do not spook. They seem interested in us, even outright sociable. Then, six canoe lengths away, they dive again. A moment later, Geri points in silence at the water just to the right of the bow. I peer into the deep blue and a torpedo of white and black zooms beneath the canoe, emerges on the other side, turns, and crosses under again. Do they wonder if we bring back the fish? I wonder if they puzzle over who took them away.

. . .

It is hot and my water bottle is empty. I lean over the gunwale to fill it with water from the lake, as is my habit, then, instinctively, hesitate. This is *acid* water, among the very most acidic lakes in this whole acid-ravaged region. My mind debates. *This is ridiculous,* declares part of my brain. *Of course this water is safe. It may kill fish but it's less acidic than lemonade. Besides, your body has a superb built-in acid-buffering system, even if this lake does not.* Another part of my brain is less sanguine. As I dip my bottle into the lake and take a big swallow, I wince, bracing for a burning sensation to erupt in my throat. None does.

The day slips quickly away. We turn the canoe west toward the portage and our camp back on Grace Lake. Quartzite ridges glisten white in the sun of a clearing sky. Now's my best chance of the day to get a photo of this enigmatic lake. We tie the canoe to a fallen tree at the shore. Geri gets comfortable with her book as I step onto white cobbles and, camera case in hand, start climbing up through a thicket of small pines and oaks to the open ridge above. Trees thin. It's hard to take root in stone.

Higher up, scattered patches of blueberries grow in thin soil in cracks in the quartzite. I grab random handfuls as I zigzag my way up the ridge. I finally reach the crest, pull out my camera, and look east down the ribbon of blue nestled between the ridges. The brilliant blue of the water and the bright white of the quartzite contrast with the deep green of the pines.

Despite a nearly clear sky above, haze, like a dirty low cloud, hunkers down between the ridges, obscuring the far end of the lake. I snap my pictures, such as they are, and retreat down the steep rocky slope to the trees and Geri below. The long portage to Grace is a breeze—all downhill.

A night of sleep in our tent, another downhill portage, and we reach the park boundary. We reload the canoe and push off into Frood Lake for our paddle back to the car. We've hardly begun when Geri points and cries out. "Darby, look at that! Water plants! Lilies! Watershield! Arrow leaf!" She sighs. "Oh, how wonderful it feels to be back on a *real* lake."

Not all lakes suffered equally in the decades-long assault from the skies. By quirks of geology and glacier behavior, some lakes, like Frood, found their watersheds supplied with minerals that could neutralize acid rain as it fell. Others, like Nellie, sandwiched tightly between the miserly quartzite ridges, quickly used up their limited acid-neutralizing capacity, then, naked, collapsed under the succeeding assault of waves of acidified rain.

We drive a hundred and twenty miles to the headquarters side of the park to visit one of the other lakes that came out somewhere between. Many campsite spaces are already occupied. The larger than usual number of bikers and walkers surprises me. We plan to paddle Johnnie Lake in the morning.

· · ·

Seven miles of tar and two miles of gravel road from camp, we unload the canoe at the Johnnie Lake landing. A bright yellow-and-brown good news, bad news sign greets us as we drive into the parking area: "No lake trout fishing on Bell, Johnnie and Carlyle lakes." Years ago there would have been no need for such a sign. Acidification had killed the trout. Acid levels now somewhat less, the trout are back. Protected from fishing, trout may reestablish viable populations.

We launch the canoe into the lake over a shallow sandy bottom that sprouts honest-to-goodness aquatic plants, an emergent sedge, and a tiny floating-leaved species of pondweed. We paddle. A gull flies across the narrow lake in front of us. More plants appear along the north shore, emergent horsetails and a narrow-leaved plant I can't identify. Loons call. We pass a small scattered clump of yellow pond lilies and a large patch of small white water lilies. I even see my favorite watershield plants and run my fingers over the sensuous mucous underside of their leaves.

Life! Green and flying and calling out. I know Johnnie Lake suffered greatly in the acid assault. But, oh what encouragement I feel here. The most obvious visible difference between the two lakes, besides the water clarity (Johnnie is very clear, but nothing close to Nellie)—no quartzite. Though Johnnie has bedrock shores, they are a beautiful rusty rose granite. Why should it matter? Different bedrocks have different composition. Different weathering rates make different lake chemistries. The lake shares one thing with Nellie. Hazy air.

Though we've seen less than half the lake, it's enough. I turn the canoe and we paddle beside gorgeous desert rose outcrops along the south shore. How can a rock exude such soothing warmth and charm? It has no sharp edges like quartzite. Its smooth worn contours seem so welcoming. A young man fishes at the parking lot as we pull off the lake.

· · ·

Friday morning. The park is abuzz. It's July 1, Canada's National Day. Cars stream through the entrance gate for the holiday weekend. We check out of our campsite, but before leaving I have questions for the park staff, and the answers come:

Yes, sulfur pollution controls have made the rain less acidic than before, but the volumes of acid-making nitrogen oxides from car exhaust remain unchanged.

Most lakes have become less acidic, and creatures are returning to many but not all. Lakes on the park's south side, exposed at least partly to the red granites, are coming back more quickly than those surrounded by quartzite. Some lakes have balked, even temporarily reacidifying. No one understands exactly why.

Biological recovery trends in other acidified lake regions of the continent are not promising. Some scientists conclude that without 75 percent additional emission reductions beyond those already achieved, habitat quality might not reach levels suitable for sensitive species. Maybe never. Some suspect global warming is hindering recovery. Cars are a culprit either way.

Then I remember. "Oh yes, what's the story on the hazy sky?" I ask.

"Smog," comes the reply. "We've had a smog advisory for several days."

"Smog? In this remote place? You must be kidding."

"It's new. Started just a few years ago. Air coming up from the Ohio Valley, Chicago, Detroit, you know, maybe Toronto too." Mike hands me a Friends of Killarney Park brochure, *Keep Killarney's Air Clean.* It urges visitors to walk and bike in the park to reduce smog-making pollution. Although the cars in Killarney create part of the problem, the brochure is blunt: "Much of this smog originated from distant sources." The smog is a visible reminder of what else the air contains.

We load our car and drive away. What face do Killarney's lakes put on the future? Is hope more than a will-o-the-wisp? That depends on what vision of "recovery" we seek for the acidified lakes here and elsewhere. The vision depends on whom you ask.

Goldeneye ducks prefer acidic lakes lacking fish that otherwise compete with them for food. "Recovery" for goldeneyes is easily achieved.

I know how the torpedo-shaped microcrustacean, *Diaptomus minutus,* would reply to such a question. *What's this about recovery? Conditions in this lake couldn't be any better than they are now.* You see, this zooplankter is blessed with an incredibly high tolerance of acidic conditions. As lake acidity rises, killing competitor species, these creatures become progressively more dominant. In the most acidic lakes, they may be the only true zooplankton species left. The entire lake belongs to them.

Would those people who work so hard to reach Nellie Lake to swim in its heavenly clear water agree with *Diaptomus*? After all, recovery of the lake in a biological sense means more algae, more plants and plankton, and less magically clear water. I do not pretend to know the depth of our collective love for lakes. I do know we will determine, by the amount of energy we consume, the composition of the raindrops that will determine the future of these lakes, these offspring of clouds and mountains.

How about those who love their lakes and love the gas-guzzling lifestyle? Would they agree with the goldeneye that, on second thought, things maybe aren't all that bad right now?

A park brochure quotes a park visitor who says he is drawn here by the "virtually untouched . . . lakes . . . [where the] power of nature reigns supreme." The words puzzle me. I must agree with the ghosts of the loons and fish and plants that once thrived in Nellie and the other damaged lakes. *Untouched? Nature reigning supreme?* Not here. Not yet. Ever?

I think about the loon pair visiting Nellie Lake to check out prospects there. The green frog waiting patiently at its edge for insects to come near. The ever-patient spider. Park users walking and biking.

Resilient. Persistent. Optimistic. Ever-seeking. Every rock, every crack an opportunity. But that's life's inherent way, isn't it. Bursting with hope. How can I be otherwise?

The quartzite isn't talking.

Limnos V—Mirror Lake,
New Hampshire

*Careful and detailed studies of individual aquatic ecosystems
are important and necessary, but researchers need to take their
blinders off! To understand a lake-ecosystem, the view must be as
large as the watershed, the airshed, the landscape and eventually
as large as the biome or planet.*

GENE LIKENS *and* F. HERBERT BORMANN, 1985

Dusk settles in by the time Geri and I finally extract ourselves from the
rush hour traffic of Montreal and catch up with southbound Interstate 89.
We cross the international border into northern Vermont in full darkness.
Headlights reveal windrows of plowed snow along the road's shoulder. How
fitting, I think, given our destination, a tiny lake in the White Mountains of
New Hampshire. Capricious weather gods choose to welcome us to northern New England with panache.

We stop at a motel in a small Vermont town, and I go in to inquire about
a room. "Only one left," the woman says.

"Midweek in late October? What gives?" I reply.

"The snow—over a foot fell yesterday. Lots of people are still without
electricity. Many came to stay here until the power's back."

Our car, with canoe on top, seems strangely out of place the next morning in the crowded motel parking lot surrounded by snow banks. I feel
embarrassed. Onlookers might question the soundness of mind of these
Minnesotans.

Snow thins as we resume travel south and completely disappears long before we reach the town of Woodstock in north-central New Hampshire. We find a small public park beside a tiny lake, nestled in the Hubbard Brook Valley, and drive in. No one is here. We spread our lunch on a picnic table two feet from our destination, Mirror Lake.

Neither the lake's diminutive size, a mere thirty-seven acres, nor any of its other features give a clue why two lake-lovers would travel halfway across the continent to visit. But those aware of the significant chapter in the life of this lake and its watershed that began over forty years ago would understand. In 1963 Gene Likens and F. Herbert Bormann from Dartmouth College began a study of the lake and its environs that would broaden and deepen our perceptual understanding of lakes. The two men did seminal research on the role of the atmosphere as a significant part of lake system relationships.

I have brought along a book, *Ecosystem Approach to Aquatic Ecology: Mirror Lake and Its Environment,* edited by Gene Likens. An aerial photograph adorns the cover, a picture worth 500 pages. Your eyes are drawn instinctively to the intense, almost blackish-blue lake that sits right of center in the picture. Glass-smooth water on one edge of the lake reflects silhouettes of shore. A patch of ruffled water runs down the middle. Two boats, mere dots, sit motionless in this footprint of a gentle breeze. Several buildings nest in small clearings near the water's edge. A mix of autumn tans and reds, oranges and yellows dapple a foreground that slopes sharply to the lake. Rusty colors from the same palette sweep off mountain ridges near the top of the photo and flow downslope to the lake. A patchwork of dark splotches overlays the rich colors, the footprints of clouds. Pastel blue sky borders the upper margin of the picture, and a tiny edge of white cloud peeks over a mountain ridge in the distance. The cut line of a freeway breaks out of trees to exit lower left. The lake sits center stage, cradled by mountains, forest, and sky.

. . .

A strong wind blows off the lake and flips one end of our tablecloth back on itself. We anchor it with a water jug and eat our peanut-butter-and-jelly sandwiches, then launch our canoe, eager to experience the lake.

We work our way against a vigorous wind. The water is clear and tinted yellow-brown. While Geri holds the bow into the wind I manage to get a

Secchi depth reading. The disc disappears eleven feet down. My reading matches that taken by researchers in October fifteen years ago, though is substantially less clear than the fourteen-year October average. According to a realtor back up the Pemigewasset River Valley, recent days have brought much rain mixed with snow. With such precipitation, much particulate material has likely washed into the lake, reducing its clarity.

Yellow leaves lie scattered across the bottom, blown into the lake by the wind. Few plants grow in the cobble and gravel shallows, though we see a few lilies, and Geri fishes out a tangle of bladderwort, a plant that eats tiny creatures like waterfleas.

We tie the canoe to a U.S. Forest Service dock beside a bedrock outcrop and begin hiking through hemlock and white pine. We walk upslope beside a brook tumbling to the lake from the direction of the freeway and discover a wooden dog-house-like structure astride the streambed; someone is sampling the stream flow. Fifty feet farther upslope sits an aluminum box on a stand, more sampling equipment. Such contraptions were at the core of Likens and Bormann's work.

An ecosystem, aquatic or terrestrial, is, at its core, a bundle of interrelationships. To understand the system you must first understand these relationships. To know a forest you study the forest. To know a lake you study the lake, but you also must study the forest, and, as Likens and Bormann saw it, you must also study what falls from the sky. They discovered that 13 percent of the phosphorus, 56 percent of the nitrogen, and 94 percent of the acidity entering the lake-forest system came in precipitation. After untold thousands of samples of soil water, stream water, lake water, rain water, snow water . . . stream flow . . . lake sediments, subsurface drainage, and chemical content of same . . . and measurements of cycles and fluxes, inputs and outputs . . . fishes and lilies, bacteria and waterfleas . . . light and temperature and wind—and two decades of time later, you *begin* to grasp the hard bottom of truth about a lake. According to ecology historian Donald Worster, "Over the next several decades the Hubbard Brook Studies became the most famous in the United States for precision of data, number of publications and comprehensiveness of approach."

Likens and Bormann perceived the lake ecosystem unit "as a watershed or drainage area, with vertical and horizontal boundaries defined functionally by biological activity and the drainage of water." We can draw it on a map. But some recent voices, some geographers, see it differently, perceiving ecosystems as dimensionless, an unmapable weaving of flows that is better represented by an organizational diagram. Ecologist James Brown argues that "ecosystems have no inherent integrity or cohesiveness as biological entities" and should be removed from the traditional hierarchy of biological organization. Differing perspectives are the grist for attaining deeper understandings.

· · ·

We return to the canoe. A heavy cloud bank has held position off to the south since we arrived at the park. It is now more prominent. The wind has subsided and we paddle easily across the lake back to the park. A man is fishing from the landing. "Catching anything?" Geri asks.

"Not yet," comes the reply. "There are rainbow trout and even browns in this lake," he continues. "You should see this place in the winter. Lots of people haul fish houses out on the ice to fish."

I ask about his faded red canoe. "It's a Merrimack," he says. "Twelve feet long and I love it." So short and squat, it looks more like a tub to me. He says he's got a cabin thirty miles away and rarely misses a weekend escape from Boston. He'd like to live here.

Another man drives up in a pickup truck. He and his wife are buying one of the new houses upslope off the northeast corner of the lake. "We live in Braintree, Mass.," he says. "Property values in Braintree are out of sight. We don't plan to live here, we just want a getaway for a few months each summer."

I leave Geri to her book and walk among birch and white pine. Dirt paths wander every which way. Tree roots laid bare by foot traffic meander along the swimming beach. A two-inch-thick windrow of white pine needles, washed shoreward by the recent rains, has accumulated at the water's edge and now holds back sand from washing into the lake.

A large birch root becomes my seat at the water's edge near the lake's outlet. The submerged remnants of an old stone dam extend to the opposite shore ten yards away, all that survives of the first dam built here in 1850 when Thoreau was completing *Walden.*

A tannery once stood here. Bark from hemlock trees provided the tannic acid required to turn hides into leather. In fact, Mirror Lake was once known as Tannery Pond. A slow but steady stream of vehicles enters and leaves this place. The feeling grows: Humans are as much a part of the lake-system as air, the watershed, or of lily plants. I feel a Limnos VI emerging. I return to Geri, and we watch a shadow creep slowly higher up the mountain across the lake, snuffing out the autumn colors as it moves, time as colors passing. The wind picks up. It is cold.

Lake of Dreams

Is it because we see things not as they are, but as we are?

ANAÏS NIN

Soon after Geri and I began married life, visions of owning our own "place at the lake" began dancing in our heads. A simple rustic cabin set close by the shore. Trees. A view of quiet water. How compelling the thought. Our search for lakeshore property began. Week after week I anxiously awaited the Sunday newspaper and the "Land for Sale" section of the want ads. We had few criteria: the lake needed to be up north, secluded if possible, and, most important, inexpensive.

After several false starts and months into our search, we learned of a lot for sale on an island in a large lake not far from the tiny mining town in which I was born. A dead-end gravel road led to a public landing on the lake beside the Timbuktu Resort. Map in hand, we paddled our canoe less than a mile to examine the property. Bound by a bedrock and boulder shore, with reasonably level land covered with pine and aspen and an asking price of $3,500, we readily agreed to buy. And so began our quixotic relationship with lake property.

Several months later we returned to our lot to make plans for our cabin's location and to identify a place for a dock. I stepped from the canoe onto a flat rock, excited to explore our property. I walked slowly into a small clearing in the trees. How grand, I thought to myself, and all this is ours!

Moments later, as I stood on a bedrock outcrop, a strange feeling, like sudden cold rain, washed over me. I looked down at the gray stone. Born of the intense heat and pressure of a restless earth, I knew this rock had rested here for perhaps 2 billion years. The lake itself had lapped this shore for perhaps 10,000 years. Even the largest of the pines around me could expect to live here for four human generations. On what basis could I, with a tenure of a few decades at most, claim to "own" a plot of ground that counted time in millennia? Though papers at the courthouse declared we owned this land, the very idea of ownership suddenly seemed ludicrous. Tenant maybe, or custodian, but surely not owner.

I continued walking and discovered a swampy spot where a cabin might otherwise have stood. We also found the land back from the lake less flat and rockier than we had realized. Where could we put a well? How would we keep sewage from seeping into the lake? How would a dock fit in among the large rocks on the shore? We returned to the landing less sure of ourselves. Doubts grew and thoughts of cabin faded. Some years later we sold the property, albeit for four times what we had paid.

. . .

But, like a persistent sales clerk, the yearning for a lake place periodically returned, and acquisition of a series of other lots followed. But each in turn revealed shortcomings after the excitement of purchase died down. Lots came and went in rhythm with the ebb and flow of our vacillations. Were we truly lake-place people? we began asking ourselves. Would we be content wedded to a single spot on a single lake? Was the associated loss of freedom to explore countless other lakes too great a price to pay?

My mother, a free-spirited woman who loved lakes more deeply than any person I have known, had a less equivocal attitude toward cabins on lakes. Inexplicably, she had no urge to have a family cabin on a lake, even rejecting the repeated urgings of her brother for our family to sign up for a ninety-nine-year lease of a state-owned lake lot. Despite the opportunity to effectively "own" a lake lot for the modest price of a lease fee, Mother never considered it, to the everlasting puzzlement of Uncle Harold.

. . .

Two years ago I learned of a pristine lake that had been recently platted. Geri and I visited the lake late that May to see it for ourselves. Of dozens of lots only a few had been built on. For Sale signs lined the road, beckoning passers-by to acquire a piece of the lake. We explored parcels by land and water. Sedges edged a gravelly cobble shore, leaving no beach, and tips of trees that had toppled into the lake reached out into the water, the submerged snags providing excellent habitat for fish and their tiny animal prey. The natural character of the lake held great appeal.

A mixed forest of oak, aspen, and conifers covered the slopes leading down to the lake. In places, steep slopes required long switchback stairways to enable lot owners to reach the water's edge. The lake's Secchi disc clarity would earn it an A+ by lake grading standards anywhere. Our canoe passed over an underwater wonderland of potamogetons and other plant species and schools of sunfish. The lake was a gem of the first order.

I began to wonder how the lake would fare as more lots were sold and more houses and docks and stairways populated the shore. I became intrigued by the opportunity to have a personal lookout point on the lake from which I could watch and record how this dream lake responded to humans over time. Geri surprised me with her enthusiasm for such a plan, and we succumbed, once again, to our on-again, off-again dream of having a place at a lake. This time, however, we compromised between our two conflicting urges. We would not build a cabin but would tent camp here instead.

One lot stood out. With more level ground near the lake than most other sites, this property would allow us a more intimate relationship with the water. The owner accepted our offer and so began what we expect to be the last chapter in the saga of Nelson lake properties and our first chapter on, let me call it Dream Lake.

Developers had removed a wide swath of brush and small trees between the lake and surveyor's flagging at the 150-foot cabin setback mark on our lot, leaving a delightful view of the water. The real estate agent we engaged to sell one of our prior lots had strongly advised us to do the same. "Lots at least partly cleared of brush," the man said, "for some reason have greater appeal to prospective buyers. If you can't do it yourself, consider hiring it done. You'll be money ahead."

Tiny sprouts of brush and aspen suckers had already begun reclaiming the clearing on our lot. My first act on the property, in fact, was to take a machete to them to keep the view-way clear. A year later, an erosion specialist informed me that brush such as I had removed supplements small ground plants to retard runoff, protecting lake water quality. I had underestimated the importance of brush.

I tried to picture what our lot would look like were I to let aspen suckers and brush reclaim the clearing. The image of a jungle of bushes and saplings did not appeal in the least. I'm convinced that the powerful allure of a golf course-like setting, our savanna-by-quiet-water landscape of the mind, is rooted in our species' very beginnings.

Dream Lake's life with intimate human neighbors has barely begun, and most of the pages of my lake journal wait to be written. As I write this after two years of ownership, more lots remain houseless than have been built on, though each year carpenter crews add a few more.

Carpenters are not building palatial estates on Dream Lake, as appear in growing numbers elsewhere. Neither are they constructing small cabins like those developed on lakes in the 1950s or before. Our owner's covenant requires a minimal building size, but no upper limit. The simple cabins of yore were designed to provide escape, respite from everyday life. Homes constructed today, and their accompanying accoutrements of the suburbs, seem more designed to bring everyday life to the lake.

Several owners of lots that slope strongly to the lake have denuded large areas around their houses, well beyond that needed for construction. Erosion experts point out that runoff water not only carries fine particles into the lake in such situations but phosphorus as well, jeopardizing water clarity. One of the owners had heard the same soil scientist that had chastened me describe the potential negative impacts of such extensive clearing, especially when conducted on a slope. The man responded defensively, derisively, challenging the validity of the expert's numbers and claiming the concerns were greatly overstated. Others on the lake have cleared extensively. One such house, one of the first built, has recently come up for sale. In over three years the denuded slope has not stabilized, and eroded runways and small alluvial fans of sand and silt show phosphorus and sediment the way to the lake, the impact of the disturbance blindingly apparent.

On one visit to Dream Lake, we met the man and his family at their shoreline. A younger man, perhaps his son, was pulling tree snags that had fallen into the water onto the shore. As he did so, he declared that, yes, he knew the DNR objected to such activity. "They say woody rip-rap benefits the lake, but I don't buy that at all," he said. The men's vehement rejection of the scientist's words bewildered me. Minnesota DNR studies do indeed unequivocally reveal such woody material plays a very important role in providing habitat for game fish fingerlings, minnows, and other aquatic life.

An aquatic scientist back home told me such rejection of experts' findings are not uncommon. As a teacher, I have long believed that information and education were the keys to understanding and enlightened behavior. Perception is more complicated than that. The men's defensiveness suggests we can fail to see what we wish not to see, that our preexisting notions may well trump hard evidence, to the detriment of lakes.

The mind may reject what we do not want to believe, but as long as water runs downhill, soil is soil, and fertilizer is fertilizer, runoff and erosion will do unwelcome things to our lakes, our denials not withstanding. Indignant at the implication that his extensive clearing might bring harm to the lake, the man down the shore declared no lakeshore owner would act against his own best interest. But, I wondered, might I act against my best interest unknowingly?

I have recently learned of another strong voice that can appear at the birthing table of perception—innate habits of mind. "Seeing" interrelationships has never been easy for laymen or scientists. Psychologist Richard Nesbett, who investigated how differently Asians and Westerners think, provides insight. Nesbett writes in his book *The Geography of Thought* that East Asians and Westerners have different cognitive structures that produce fundamentally different ways of seeing the world. He opens with a remark his Chinese student Kaiping Peng made to him. "You know, the difference between you and me is that I [as an Asian] think the world is a circle, and you [as a Westerner] think it's a line."

Our mind-sets inevitably determine how we see, interpret, and understand the world. Nesbett believes that while we Westerners tend to see the world as a collection of objects, typically separate from their context, East Asians have a "holistic view focusing on continuities in substances and relationships in the environment." Consider two examples from his book.

When people were shown underwater animated scenes containing a large, fast-moving fish plus plants, stones, bubbles, and smaller slower animals and then later asked to describe what they had seen, the initial response of Japanese was typically something like, "It looked like a pond." In other words they were struck by a setting, a context. Americans typically responded, "There was a big fish, maybe a trout, moving fast," appearing to be taken with an object instead of a broader setting.

In another experiment, children were presented a picture of a cow, a chicken, and grass and asked which of the other two goes with the cow. American children grouped the chicken with the cow, seemingly because they both were animals, a taxonomic category. Chinese children responded differently. I tried this out on my friend Na when she and her family shared Thanksgiving with us. She was born and raised in China and came to the United States in her twenties. "Well," she said, "a Chinese person would match the grass with the cow because that's a relationship—cows eat grass. Americans would do the other." I also asked for her reaction to Kaiping Peng's observation: that Chinese search for relationships, that they feel a need to understand the whole before understanding the parts, whereas Americans focus on objects and categories. "Yes. Yes. Yes," she exclaimed, delighted that others saw the differences in thought patterns long familiar to her.

The Chinese tendency to look to the whole for understanding predisposed them to be more perceptive of explanations of phenomena that involved "action at a distance" than Westerners. Such a way of thinking enabled the ancient Chinese, for example, to believe that ocean tides were caused by "movement of the moon," ages before Westerners saw it that way. We long believed falling objects fell because the *object itself* has the "property of gravity," instead of surmising that an external force acted upon it.

How much sooner might we have understood lake as an interconnected system, molded by watersheds and airsheds, had we a cognitive predisposition to perceive context and relationship and the potential for action at a distance? The man down the lake is not unique in his thinking. Our tendency to see objects before relationships impedes our perception of the true nature of lakes.

. . .

Geri and I return to our land several times a year, and I fill more pages in my field journal. So far I detect no discernible change in the lake since we first saw it, though change is not always readily apparent. Homeowners are, by and large, abiding by the proscription against clearing land within fifty feet of the shore, except for access space to get down to a dock. Many owners have limited the cleared width of their paths to the water to an absolute minimum, leaving maximum width of their lake frontage clothed in the trees and shrubs that were originally present.

I think about the long-term prospects for Dream Lake. In its favor, the lake's watershed is relatively small and its maximum depth is over six times greater than that of Diamond Lake. Its steep slopes, however, put it at notably higher risk of runoff of silt and phosphorus and nitrogen. Human factors are less knowable. As houses continue to appear, how many will bring with them the landscape of suburbia, the ancient savanna landscape of the mind, with lawns and mowing and fertilizers and removal of "unsightly" tree snags and pond weeds from their shores?

. . .

Lakes have limited life spans. All will someday die, will disappear because of changes in drainage or climate or geologic events or through fill-in of their basins. An erosion expert has estimated that Dream Lake, left alone in its natural forested watershed, could expect to "live" several tens of thousands of years more. Its actual life span, however, depends directly on the perceptions of those of us who come to its shore. He went on to say that if we pursue development, "business as usual," turning lots into golf-course savannas down to the water's edge, the lake's life could be reduced to little more than a century. Enough current residents have forgone business-as-usual-development so far that the lake's actual life span is likely to exceed another hundred years. How much longer depends.

Henry's Mirror

Walden is not significant as a place at all, . . . It is significant only because the word Walden suggests some thoughts a man had once. Where he had them really doesn't matter.

RAYMOND ADAMS

No parallel tract or body of water or place has so captivated the human imagination or so taught us how to relate to the natural world from which we spring. Nowhere else is there a . . . monument of such breathtaking consequence. . . . Walden is an international shrine.

JANE HOLTZ KAY

At a small bridge nine miles west of Boston a minuteman fired a shot heard "round the world" in 1775. Five miles west and seventy-eight years later, Henry David Thoreau fired his own shot beside a small lake physically indistinguishable from hundreds of others sparkling in the Massachusetts landscape. Echoes of the revolutions ignited by both shots reverberate still.

As Thoreau is our most famous lake-watcher, his beloved Walden Pond is surely our most famous lake. Seven hundred thousand people annually visit that body of water, offspring of a block of ice abandoned by a wasting glacier. Some come to fish, many to swim or saunter along the paths that surround it. Others come to make peace. Some come on a pilgrimage as to a holy place. I have read that Gandhi dearly wanted to visit Walden.

This is my second visit. I have longed for years to measure the clarity of Thoreau's "clear and deep green well . . . water so transparent that the bottom can easily be discerned at a depth of twenty-five or thirty feet." Is it still so? How has time treated Henry's beloved pond?

I drop off Geri and the canoe at the boat landing on this glorious October morning and leave to park the car. I insert my ten dollar bill into the fee machine at the parking lot gate, collect the five Susan B. Anthonys that tumble out as my change, and join a mere handful of cars in the large lot. I do not imagine it is this easy when the hordes of summer visitors flood these grounds. I follow a dirt path, cross a narrow highway, and descend a wooded slope toward scattered patches of blue in a mosaic of yellow, orange, and crimson leaves, radiant in the bright sun.

As my steps bring me nearer the water's edge, a sudden rush of emotion wells up within me, startling me. Is it the lake, this setting, Thoreau's mystic that touches me so deeply? My pace slows as this powerful landscape of mind settles over me.

In *Walden,* Thoreau writes in "The Ponds" chapter, "The scenery of Walden is on a humble scale, and, though very beautiful, does not approach to grandeur, nor can it much concern one who has not long frequented it or lived by its shore." Henry, I must vehemently disagree! I have "frequented" this lake, as you put it, only once before, fleetingly, when Geri and I brought our two children east for a squeeze-in-all-the-stops-you-can American history tour. We tarried here no more than two hours, never swam nor fished nor walked the pond's full perimeter, then rushed off to Lexington and that famous bridge. Though I'm unable to fully explain this power of place I'm feeling, Henry, your lake clearly "concerns" me.

The path on the south shore takes me to Geri and our canoe. The lake is the mirror Thoreau wrote about, reflecting "the bright tints of October." Unusually heavy rains this autumn have delayed leaf fall several weeks. A freshly dead rainbow trout lies in the sand at the boat launch.

We paddle from the landing, veer west, and move slowly along the shore in a canoeist's version of an idle stroll, aquatic sauntering. A silent motor pushes a lone fisherman in his blue boat slowly down the middle of the lake. A trout rises, dimpling the glass surface. The morning sun emblazons the west shore, and the bright crimson of a brash young maple, like an arboreal chanticleer, shouts a silent HUZZAH at the prospects of the day. Though Walden is not large, sixty acres in all, numerous small coves sculpt an engaging shoreline.

Only traffic sounds from the highway and a distant drone of an industrial machine, white noise, blur the quiet. The earthy smell of the woods, of dry leaves and leaf mold and humus overwhelms any smell of lake.

We enter Little Cove as the call of a train, invisible behind a wooded point, grows louder. The clackity-clack of its wheels reach the far end of the lake, now drops in pitch, and fades to the west. The Fitchburg line has passed this way since Thoreau's day, when farmers set their clocks by the train whistle.

Another train. Not ten minutes have passed since the last. This time we watch silver gray commuter cars rush east toward Boston. We reach Long Cove, where trains pass closest to the lake, the rails not twenty yards from the water's edge. Trainmen who looked out over these crystal waters in Thoreau's day developed their own relationship with the lake. One suggested renaming it God's Drop.

Large stones line stretches of the shore as though bulwarks against the pressing land. Occasional boulder-lined runways pave tiny gullies, preventing runoff water from carrying soil and sand in a headlong rush to the lake.

A young boy's voice calls "Mommy" from back in the woods in the direction of the trail that circles the lake. The trail occasionally reveals itself in openings in the trees. A man walks past, then a couple speaking Spanish, two more men saunter by, then a woman pushing a baby stroller and leading a little boy. A minute later a man in running shorts follows, power walking, his shirt dripping with sweat.

A longer gap in the trees reveals two wire fences and steel posts running on each side of the trail. These parallel fences confine hikers to a single-file path, strikingly similar to cattle fences I've seen farmers use to guide beasts from one field to another. Here the fences prevent people from leaving the trail for a slide down the bank to the water. Rectangular white signs attached to fence posts appear every so often, reinforcing the statement of the fence.

Surely Thoreau would rebel at such restrictions. My instincts are with Henry, but I understand the purpose of the fences. I remember vividly the severe bank erosion we saw our first time here, before there were fences. The trampling feet of hordes of visitors had eliminated ground plants, and soil compaction had killed many trees. Few roots remained to hold back the soil. I think Henry would be no less disgusted at such erosion-driven degradation of his pond than with these fences.

Thoreau had his own apprehension about the future of the pond and its surroundings. He feared ultimately the land would become private, and "fences shall be multiplied, and mantraps and other engines invented to confine men to the public road, and walking on the surface of God's earth shall be construed to mean trespassing on some gentlemen's grounds." Walden and its immediate vicinity are now publicly owned.

An official opening occasionally breaches the fence on the lake side of the trail, allowing a walker to step down a short flight of boulders to cool her toes in the water.

We leave Ice Fort Cove and, by the time we approach Thoreau's Cove, hear more Spanish, Indian, French, and Asian tongues wafting out from the trail. A rotund man with a partly artificial leg emerges from bushes onto a stretch of sand at the mouth of Thoreau's Cove with a folding lawn chair, backpack, and fishing rod. Geri asks what he's fishing for. "Rainbows," he replies. "They stock rainbows in the lake. You can legally take three, but I always let them go."

Three fishing boats now work the waters of the pond. None of them make even a whisper as they move along. Gasoline motors must be forbidden.

I've still not sniffed the familiar smell of lake. The near-complete lack of plants decaying at the shore may partly explain its absence. The smells of white pine and pitch pine mixed with the more subtle odors of earth mask any nascent lake smells.

We reach the main swimming beach and pull the canoe onto the sand. No one swims in the cold water. Down the beach a small group of people have gathered around a deeply tanned elderly man in a bathing suit sunning himself in a lawn chair. Curiosity draws us to them. The eighty-eight-year-old claims he has swum here every day for the last twenty-three years. He stops only when the winter ice becomes too thick for him to break with a club. He complains that park authorities won't let him employ more sophisticated technology like an axe or saw. Bill is a gregarious fellow, the Grand Old Man of Walden's Red Cross Beach. He introduces us to his beach friends as they come and go: fellow senior swimmers nearly as fanatical about swimming here as he, and an elderly woman who has just published a book on crows.

Visitors wander off, giving me an opening. "You meet a lot of people here, Bill. Why do they come?"

"Oh, lots of reasons, I guess. Many swim. Fishermen come for the trout. For others this place seems to have a soothing effect, particularly for people who've had major stress in their lives—divorce, death of a child, nervous breakdown, heart attack, that sort of thing. There is something very calming about this place."

Bill notices an elderly woman has begun swimming, and he breaks off the conversation. "Time to get into the water," he says, as he gets out of his chair and removes his shirt. I watch him enter the water, skeptical he will do more than a few strokes before being driven ashore by cold water and age. I am wrong. His smooth relaxed stroke carries him farther out from shore than I expect, then back and forth, paralleling the beach, comfortable as a porpoise. I watch and wonder. What compels him to come here every day to swim? Is he emulating Thoreau, who frequently took morning swims? Is it something else? I think of Brett Hesla's lyrics in his song "Lake of Mercy."

> *From the shore, from the shore,*
> *   where moonlight bathes the sand,*
> *I set sail, I set sail and leave behind the land.*
> *Let me rest upon your waters,*
> *   Lake of Mercy, precious friend.*
> *Rock me softly in the midnight wind*
> *   and make me whole again.*

After ten minutes I realize we must get back out on the water ourselves if we are to get a Secchi depth reading under noonday sun. We steady the canoe over the deepest part of the lake and the disc descends. It disappears just short of sixteen feet down, half the depth of visibility Thoreau claimed. What has happened? The heavy rains earlier would have increased runoff, temporarily decreasing water clarity. But this much change? It doesn't seem likely.

Professor Secchi's disc had not been invented when Thoreau watched fishes at such great depth, so I cannot precisely compare the lake's clarity then to now. However, Henry was an acute observer and a surveyor. I doubt Thoreau's estimated depth of clarity was far wrong.

This water today is still attractively clear compared to most other lakes across the country. It would receive an A grade by clarity standards in the Twin Cities. But I expect Thoreau would detect in an instant that this "gem of the first water which Concord wears in her corona" has lost some of its luster. U.S. Geological Survey investigators report the lake indeed has suffered significant eutrophication over the past seventy-five years. Thoreau's words that Walden "needs no fence" suggest he never countenanced such degradation. If Henry David Thoreau couldn't see lake degradation coming, perhaps we can be forgiven for not seeing it either.

· · ·

Allowed to accelerate unattended, eutrophication's consequences include population explosions of those nefarious blue-green algae, a loss of life-sustaining oxygen in the lake's mid-depth waters, a change in the species composition of a lake, and even fish kills—the kinds of things that trouble other lakes in Massachusetts and elsewhere.

Though researchers document significant eutrophication here, the most recent studies reveal this unfortunate process is, for the moment, stabilized. That stability appears to depend on the well-being of a fascinating form of life few visitors see in this lake—stoneworts.

These moniker-challenged beings are a kind of green algae, though not the specks of green you may have seen in lake plankton, nor the hairlike green strands attached to submerged rocks. Stoneworts are particularly large for freshwater algae. The stonewort found in Walden, *Nitella,* can grow as long as your shin bone.

Like its close relative *Chara,* and many kinds of aquatic plants, *Nitella* has a spindly stem that looks like a starved miniaturized version of those long skinny balloons that clever people twist into giraffes and other animals to delight children. Imagine a bright-green one twisted off at intervals to form a series of elongate segments that you'd instinctively call a "stem." Now imagine a cluster of short skinny tubes sprouting from each of the twist points that you'd instinctively call "branches." Fresh, *Nitella's* stems and branches look to be made of beautiful green translucent Jell-O. Now imagine each segment is no more than two inches long and of pencil lead diameter and you've got an idea of what an individual *Nitella* looks like.

These plantlike beings are eminently worthy of such detailed attention. Without them, the ecological stability and aesthetic appeal of Walden Pond would likely collapse in a heartbeat. *Nitella* is the keystone species to an intricate web of interconnections in the lake that would make a spider dizzy.

Like Lake Itasca and many temperate lakes in summer, Walden has an upper layer of warmer and well-oxygenated water, a transition zone of dropping temperature and oxygen levels, and a cold oxygenless bottom stratum.

I'm told *Nitella* grows in vast underwater meadows on lake bottom that lies from twenty to forty-five feet below the surface. There it uses all the dim light it can capture for photosynthesis. This process releases oxygen. *Nitella* also extracts much phosphorus from lake water to perform its body chemistry.

Long before there were university departments of limnology, and fifty years before the father of limnology, Edward Birge at the University of Wisconsin, studied and coined names for these three summer lake layers, Henry Thoreau discovered the layering phenomenon for himself. Though crude, his method revealed the heart of the matter. He lowered a bottle of surface water fifty feet and let it rest there for thirty minutes to equilibrate. Then he retrieved it to the surface and took its temperature. He repeated the process at one hundred feet. He compared the temperature at the surface and the two deeper measurements and calculated that the water had cooled at a rate of one degree for each five feet of depth. (Had he taken samples at other depths also, he would have discovered the cooling rate was actually not so uniform.) Curiosity—the driver of science, the mother of "seeing."

The middle layer, and the trout that love it, are currently blessed with an oxygen supply through the challenging summer period, at least for now—thanks to photosynthesis by Walden's vast meadows of stoneworts.

Trout can never realize that those acres of spindly *Nitella* are doing something else for the lake as important as making oxygen for animals. It has to do with that driver of eutrophication, the alchemist that can turn clear lakes into a blue-green algae soup—phosphorus.

Blue-greens thrive in phosphorus-rich waters. By retaining large amounts of phosphorus in its cells, the large stands of *Nitella* prevent blue-greens from getting it. The oxygen *Nitella* produces further stymies the blue-green's quest for phosphorus by preventing phosphorous release from bottom sediments, where much is chemically bound, putting the blue-greens on a starvation diet.

So, all seems well with Walden as long as the *Nitella* carpet remains robust. Unfortunately, some *Nitella* species don't fare well in eutrophic waters, raising the question, how stable is the current condition of the lake? That is unclear. An uptick in phosphorous that damaged *Nitella* might spin everything out of control. More blue-greens mean murkier water and reduced light penetration to the remaining *Nitella,* pushing the lake into a downward spiral and ultimately producing ever more blue-greens, more murk, and *very* unhappy humans.

Lowering phosphorus input to the lake would likely lower the chance of disrupting the equilibrium. Phosphorus is the key. Phosphorus gets into the lake in several ways. The restrictive trail fences and boulders protecting the shore banks reduce phosphorus input from runoff. But I think about the many colored leaves scattered on the lake bottom near Ice Fort Cove. Each carried its own tiny load of phosphorus into the lake from the land.

I think also of that dead trout at the landing. In decay, it adds phosphorus to the lake, though a fish taken home to the grill takes some away. Visiting water fowl would leave small amounts as well.

U.S. Geological Survey studies find these amounts are minimal compared to other sources. Their summer month numbers show that significant amounts of phosphorus enter the lake from groundwater. Even more phosphorus comes from dust and other dry airborne particles that fall from the skies. The greatest source, making up over half the summer phosphorus input to Walden? Swimmers. Swimmers? Ah . . . er . . . , oh yes, swimmers.

The ancient alchemists, who tried to turn cooked, putrefied urine into gold but discovered the element phosphorus instead, could have warned us. My mother could have told as much, as well. I remember her standard reply to my sister or me when as youngsters playing at the lake we'd rush up to her whispering, "I have to go to the bathroom."

"Go into the lake," she whispered back.

Apparently other mothers give similar advice. If only we could train bladders, adults' and kids' alike, to forgo that instinctive urge to empty when we're in the intimate presence of bodies of water.

Reductions in phosphorus-laden dust falling to earth or flowing in by groundwater are long-term strategies. Reducing human pee could be another matter. Researchers euphemistically concluded that a "swimmer education program" is the most reasonable approach to reducing phosphorus input into Walden to protect the all-important stoneworts and therefore Walden's purity.

So it all comes down to pee. Who'd ever have thought it? Imagine Walden's beach front lined with Porta-Pottys, bright yellow ones at that, emblazoned with "Pee before entering the water," or "Pee first." What would Henry think?

· · ·

The western sun now nudges shadows out of the woods onto Red Cross Beach. I cannot guess Walden's future. Though Thoreau's pond and woods are "owned" by the people of the world, it has fallen to Massachusetts to serve as caretaker. I do know the administrators of Walden Pond State Reserve face conflicting mindscapes of the pond's future among its adoring public. Some argue visitor numbers should be dramatically reduced, that the pond and its surroundings be left as Henry left it. Others disagree. Restrict or deny swimmers' or hikers' access? Not on your life! The number of parking spaces is a valve on visitor pressure now. Threshold points on valves can be changed—in either direction.

No one has consulted *Nitella* or the trout.

Henry's "shot" gave birth to an environmental ethic, to a sensitivity to matters of humans and place. But it is not a road map. What is a "proper" presence for humans in a natural landscape? Thoreau offers little help reconciling the varied visions of lakescapes that live in our minds. We must sort out these mental landscapes for ourselves.

Brilliant scarlet leaves blaze in a line of brightest fire at Main Beach as we land the canoe at the boat landing in the afternoon sun. The lump in my throat returns as I walk the trail to the car.

Lakescapes of the Mind

*There are innumerable avenues to a perception of truth. . . . It is
not in vain that the mind turns aside this way or that: follow its
leading; apply it whither it inclines. Probe the universe in a
myriad points.*

HENRY DAVID THOREAU

Though infinitely more famous than the rest of Massachusetts' lakes,
Walden Pond was not the only lake in Henry Thoreau's life. Four others,
sprinkled about the landscape of Concord and Lincoln, made up the rest
of what Thoreau called his "lake district." One of those lakes, Flint's Pond,
Sandy Pond to some, stands apart from the rest.

Thoreau was familiar with Flint's Pond long before he lived his experiment at Walden. His roommate at Harvard, Charles Wheeler, lived a short
distance away from the pond and had built a cabin on its shore as a summer
retreat, perhaps with Thoreau's help. Henry apparently spent six weeks
there with Charles the summer after finishing college. Some think that experience sparked the idea to build his famous shanty on Walden years later.

. . .

Flint's Pond, the only nutrient-rich lake Henry writes of, is also the only
lake he denigrates. Thoreau obviously preferred clear-water lakes. He notes
that Flint's pond lies at higher elevation than Walden, and a small chain of
ponds between them hints the former may once have drained into the latter.

Thoreau invokes God's forbiddance that such flowage should ever return to mingle the impure water of Flint's with the pristine water of Walden.

Thoreau had strong opinions about Farmer Flint, whose land bordered the lake and who had given the pond its name. Thoreau was incensed with the man: "What right had the unclean and stupid farmer, whose farm abutted [the lake] . . . whose shores he ruthlessly laid bare to give his name to it . . . a skin flint who loved better . . . a dollar . . . who regarded even the wild ducks which settled in [the lake] trespassers. . . . Fingers grown into crooked and horny talons from the long habit of grasping harpy-like . . . who would carry the landscape, who would carry his God, to market if he could get anything for him."

I can only wonder what drove Thoreau to write with such venom. Flint's farming has desecrated the lake's shore in Henry's eyes, but his wrath is inconsistent. He acknowledges that woodcutters, the railroad, and he himself have profaned Walden, yet the pond remains unchanged. Henry condemns Flint but absolves himself and others for their own lake transgressions.

· · ·

More than 150 years have passed since Thoreau wrote of Flint's pond. Do houses and docks and lush green lawns now crowd the water's edge? Do cattle muddy its shores? Is the pond covered with scum? Choked with pondweeds? Is there even public access to its waters? Now that we are here in Thoreau country, I can't wait to find out.

We inquire at the Old Lincoln Town Hall for directions. After several wrong turns on narrow New England roads we approach a lake. No sign tells its name. A short driveway leads to a tiny parking lot at its edge. The water body is the size and shape of Flint's Pond on our map. A small pump house sits partly in the water. A blocky gray building borders the parking lot. The lot is empty.

I dash to the front of the car and begin untying canoe ropes. Then I see it—a white rectangular sign, stout and sober, at the edge of the tar, the blue lake behind it.

"Absolutely No Swimming, No Fishing, No Boating. By order of the Lincoln Water Commission."

I turn to Geri. "So what do we do now?"

I want desperately to get onto that lake. I don't need much time there. I want only to learn its condition, take notes on aquatic plant life and the shore, and measure the water clarity.

"But it doesn't say anything about canoes," I say. "I know of lakes where motorboats aren't allowed but canoes are." Geri says nothing.

"If they really don't want *anybody* on that lake for *anything,* they could just say NO TRESPASSING. That would take care of it. But they didn't. I bet a canoe is OK." Geri looks at me soberly and shakes her head.

We look at the lake, the canoe, and again at the sign. "Maybe if we explained that we need less than an hour to do our research they'd give us permission," Geri finally says. "The woman at the Town Hall told us there was a Water Department building just down the road. Let's see what they say."

We return to the road and notice a sign urging cautious driving, that little salt is used on this road in winter, apparently to minimize pollution of the lake. Several hundred yards down the road we find the Water Department building and walk in. A woman listens politely to my plea, then says, "Access is normally not allowed. Anyway, we don't have the authority here to give you permission to go on the lake. You'll have to talk to Pat Allen. His office is back at the building by the lake. He was probably on coffee break when you were there. He'll return soon." A coworker across the lab rolls his eyes and smirks as she speaks, as though to say the chances of getting approval are about as good as a trout surviving an evening in a hot tub.

Not ten minutes after we return to the parking lot by the lake, a pickup truck drives in and two men get out. One wears a blue work shirt with "Kevin" embroidered on the front. The other man, in dark work clothes, must be Pat Allen. They look in silence at us and our canoe. I introduce myself and Geri and explain why we're here. "I just need a short time on the water," I plead. "Long enough to take notes on the aquatic plants and a Secchi depth measurement. Could we be allowed onto the lake for just a short time, please?"

The men look at each other in silence, then Mr. Allen asks, in a delightful Irish brogue, "Research? You want to do research?"

"Yes," I reply.

More silence, then Kevin, looking at his boss says, "I guess it's all right." I profusely thank them and providence alike.

Now they become talkative about their lake. It's a water supply for the town of Lincoln. "Concord used to pipe water to town, too," Kevin says, but "years ago we kicked them out."

I tell them Thoreau seemed to think less of this lake than the others, how he described its waters as impure. Kevin is quick to reply. "Actually, people around here will tell you this was Thoreau's favorite lake. He used to come here often. He had a favorite rock he liked to sit on. See that island just off the west shore? Thoreau's rock is straight at the shore from that island. Look for it when you paddle past. It's big. You can't miss it. In fact, he'd have built his cabin here if old man Flint had let him."

Kevin's words bring me up short. Thoreau would have built his cabin here? But Thoreau felt Flint had desecrated the place to a fare-thee-well. Quite the perceptual contortion, Henry, I think to myself.

Kevin points to a small aluminum boat lying on the grass just above the crushed rock shore. "That boat's not for emergencies," he says. "We've had goose problems. At times, geese were so thick you could practically walk to the island on their backs. I shoot fireworks 'bullets' at them to chase them away. When that doesn't work, I put the motor on that boat and buzz around and around among them. They hate it and finally just give up and go away. We had to get special permission from the state to harass them."

The men have work to do, and they get back into the pickup and drive off, leaving Geri and me alone with Flint's Pond.

I am untying canoe ropes as Geri is gathering life preservers, paddles, field journal, and the Secchi disc when a shout comes from a man jogging with his dog up on the road. "What do you think you're doing!" he yells. "You can't put that canoe in this lake!"

"We've got permission," Geri yells back. The man turns in his tracks and trots down to the parking lot toward us. "We've got permission," I say calmly.

"Permission from whom?" comes the stern reply.

"Pat Allen."

Lines on the man's face soften. "That's the name I had to hear." I explain what we wish to do here. "I'm Andrew Hall. I've been on the Town Water Commission for fifteen years, and I've never known Pat to say yes."

We look at the lake. I see no development, not even farms appear anywhere on its shore. "Much of the land is owned by the town, some by the state, or is under conservation easement to protect the lake," Andrew says. "There are three houses on the lake, just out of view down the shore to the right. All were built in the early 1900s. They've moved their septic systems to the front of the houses to make it harder for phosphorus and nitrogen to seep into the lake. They're not allowed to fertilize their grass."

Mr. Hall leaves. The canoe ready, we launch this warm November 1, under mostly clear skies, and begin a counterclockwise circumnavigation of the shore. The water is clear but tinged with brown. Though more colored, the water lacks that wisp of murkiness I noticed on Walden yesterday.

Scattered white pine poke green heads above a mostly yellow and brown oak forest. The water reflects the bright blue of the sky. This lake is a gem. Things must have been very different in Thoreau's day.

We paddle close to shore. I'm looking for aquatic plants. This late in the year I expect them to be senescent, leaves brown and frayed and stems torn. I see none. A lone black cormorant hunkers low in the water far out in the lake. Three ducks explode out of a tiny cove. Rapid wing beats move them quickly across the water. Not a goose anywhere. We pass the three old houses all back among the trees on a hill. I wonder what prompted builders a century ago to place houses such respectful distances back from the water.

A small marshy area with tangled sedges appears a third of the way up the east shore. This must be a source of the water's brown tint. Still no emergent or floating leaved plants. Small darkish plants hug the bottom on patches of sand. A pop can's aluminum shine reflects from the gravel bottom. I draw-stroke the stern against the wind until I can pull the can from the water. It looks to have been in the lake a long time.

A second wetland, larger than the first, dominates the lake's northeast corner. A black organic layer, maybe eight inches thick, rests on sand. Waves slowly eat away the black. A swarm of tiny mayflies suddenly appears. Our words of delight startle a sunbather lounging on the concrete foundation of another pump house. His towel and shirtless torso melt into the brush in haste. Is it even illegal to sun oneself here? Maybe he rushes off to report our presence to the police.

Tree branches and shrubs overhang the water everywhere. Thoreau used to come "a-chestnutting" here on windy fall days when the nuts dropped into the water and washed up by his feet. The ripple marks he admired in the sand still cover the bottom, but the ranks of rushes that "grew in waving lines . . . rank behind rank, as if the waves had planted them" are gone. Other plants grow in the sand. One has a rosette structure, another fine needle-shaped leaves.

A kingfisher swoops from branch to branch, leading us down the west shore toward the waterworks buildings as three tiny ducks floating tightly together catch Geri's eye. Their heads seem too large for their little bodies. One submerges. The others instantly follow. One pops up and so do the others, like children playing follow-the-leader. Their markings say bufflehead.

We approach the island and scan the shore for Henry's rock. Several large stones hide behind the brush and young trees that skirt the water's edge. There's one that might be Henry's.

The noon sun reminds me we must head to deeper water and take a Secchi disc reading. Away from the protective shore, erratic wind gusts buffet the canoe. We hold position until a patch of calm moves our way. I quickly lower the disc. At fourteen and a half feet the white disappears. Can the water really be that clear? The Secchi disc doesn't lie. How could Thoreau have disparaged this lake?

Andrew awaits us as we pull the canoe onto shore at the Water Department site. "What a lovely lake," I exclaim. "What an example of how a town can protect its water. So encouraging. Back home, Minneapolis takes some of its water from the Mississippi. Sure are no comparable protections like this upriver. The closest thing I've seen to what you are doing was in Norway. Oslo takes drinking water from a lake, too, and they also vigorously protect its quality. What a message of hope. I wish people back home could see this. How did you manage to pull it off?"

"We've got a three-member Town Water Commission. We're elected by the people, but we don't answer to the selectmen of the Town Board. We have our own funding and retain unspent funds from year to year. As a result we are more removed from politics than the selectmen could be when it comes to protecting our water. We have authority to stop all lawn watering if we need to.

"I've been thinking how else I could help you. Maybe you can use this groundwater protection study done for the town years ago," he says and hands me a thick booklet.

He must leave, he says, then lingers and wistfully avers, "In all my years on the commission I have never been on the lake. Here's my email address if you have more questions," he adds. "Leave the report with Pat when you're finished." I take the booklet and begin writing notes.

Years ago groundwater contamination forced six western Boston suburbs to close city wells. Three of these communities border the town of Lincoln. No wonder Lincoln is so protective of Flint's Pond.

The lake's watershed is only twice the area of the lake's surface. Control land use immediately adjacent to lakeshore in such a situation and you control much of the potential pollution.

Pat emerges from the larger building and joins us at the water's edge. "Well, what did you learn out there?" he asks.

"It's so different from the impression I got from Thoreau's writing. My Secchi disc reading was over fourteen and a half feet. Walden was not that much clearer yesterday. You can't call this water impure. It's not choked with weeds."

Apparently, I sound like I'm questioning the validity of Thoreau's observations. "What Thoreau wrote might have been true when he was here," Pat reminds me. "We have old photographs of cows standing in the water, even stone fences going a distance out into the lake. You know what that meant." I nod.

"Water often becomes more clear in late fall," I say. "But even so, the lake's appearance is impressive."

"Yes, we have algae blooms in August, but things don't get terribly bad," Pat says. They have apparently used an aquatic herbicide in limited places on the lake in the past.

I show him the report Andrew loaned me. "Let's go to my office," he says, "I'll make copies of whatever pages you need."

Pat originally came from Ireland and got started in the plumbing and heating trade in Boston. Then he came to Lincoln. He's been at his Water Department job for thirty-two years, and obviously loves his work. Copying finished, he gives me his phone number and says to call if I have questions. I thank him again for the high privilege he has granted me this day and walk outside.

Thoughts tumble and swirl in my head unrestrained. Beauty. Thoreau's rock. Secchi numbers. Lincoln Town's Irishman. A commissioner's passion. Landscapes in conflict. Farmers on Diamond Lake. Ripples in the sand. The images need time to dance in the landscape of my mind.

What thoughts might pass through Thoreau's mind if he could see this pond now? I feel compelled to speak to him, and return to the shore.

Henry, how I wish you could break the bonds of time and stand beside me this lovely day so we could talk. I'm at the south shore cove no more than three hundred yards from that favorite rock of yours. Seen from here this hour, Flint's Pond is stunning, a piece of heaven's blue snuggled into the landscape. What a wonderful thing the passage of time and the citizens of Lincoln have wrought.

Except for a small meadow between your rock and me, neither house nor barn nor field nor any contrivance of man mars the beauty of this place now. Flint and the farmers and would-be land developers that might have followed have been vanquished from this place. I sense you smiling.

Even the less pure waters you describe for this pond have been vanquished. The water here is almost as clear as Walden's. I feel your joy at this news. I know your lakeside pastimes and anticipate your next thought: When do we go fishing? I'm sorry, Henry. We can't. Not that there are no fish, the cormorant attests to their presence. It's just that . . . well . . . fishing is forbidden. About your passion for swimming? That's forbidden too. Before you look for your favorite lakeside trees to gather chestnuts, I must tell you that blight vanquished chestnuts from North American forests long ago. But we could not gather nuts at the shore even if the trees were still here. You see, we are forbidden from walking within fifty feet of the water's edge.

Your silent thoughts thunder in my ear. *Outrageous!,* you say. This too steep a price to pay for clean water? Would you rather revert to the old days, the old ways, Henry? Your vitriolic verbal assault on Old Man Flint's despoliation of his land by the lake *screams* NO! Clearly Flint's vision of lakescape differed from yours and, at least when I began this journey, from mine. I think back to the delight that filled my being mere hours ago when I looked out on this pond for the first time. Beautiful water, ducks on the wing, the lake edged with the glorious colors of fall, and not a single artifact of humankind to disturb the notion that this was lake primeval, lake like God intended it to be, lake in its true essence. It seems to me that you and I see the "proper" condition of lake as nature undisturbed.

. . .

I set out on my journey seeking an explanation of the paradox of why, despite our professed collective love for lakes, we help speed their deterioration. Since how we "see" a thing determines what we do with that thing, the explanation of the paradox appeared to lie in a failure of sensory input. Hindered by the cloak of invisibility that prevents our directly seeing or experiencing for ourselves the marvelous beings in their homes beneath the waves, we understandably stumble at the edge of stewardship. Could everyone fully see and understand this fantastic world of lake, stewardship would spring alive like a thousand water lilies bursting into bloom.

At least that was what I once thought. Henry, like you, I have my own lake district back home. While Flint's Pond may be nutrient-rich, my Diamond Lake is excessively, pathologically, so. Part of its degraded condition is attributable to the several farms along its shore. As I explored the lake, I was angered by cornrows coming to within a paddle's reach of cattails at the water's edge, allowing for no mitigation of excessive runoff of nutrients and herbicides directly into the lake. I was exasperated that cows freely trampled the bank eroding soil, dropped cow pies on slopes mere yards from the lake, and stood knee deep in the lake, gushing urine directly into the water. The disgusting pea soup color of the lake water depressed me.

In the course of my explorations on Diamond, I visited individually with three octogenarians whose farm family tenures by the lake go back to first settlement, about the time *Walden* was published. Two of the men were born and raised on the shore of the lake. The third farmed his entire life close by. The two things I most wanted to learn from them were how clear the water was in the earliest of family memory, and when the lake made its fateful turn into its current hypereutrophic state.

I learned much else. Of families gathered together at the Kimball place in fear of Indian raids. Menfolk going off to fight the Civil War. Droughts and farming on an almost totally dry lake bottom. Of bountiful crops of potatoes and corn from the rich lake bottom soils. Drownings and a suicide in the lake. The return of wet years and cattails and a lake full of muskrats.

Of clouds of red-winged blackbirds rising from the cattails to feed on oats and corn, and shooting and pot banging to drive them away. Duck hunting on the lake as far back as anyone can remember. Of minnow dealer's traps along the south shore.

"But what was the lake water like," I asked Hilmer.

"I don't know. Never paid attention to it. Too much work to do on the farm."

Arnold's answer: "It was a mud hole."

"Vern," I asked, "was the lake important to you and your family?"

"The lake was always significant to us," he replied.

"How so?" I asked. Now, I thought, I'll hear stories of golden water mirroring golden sunsets, the soul-touching joys of life growing up mere steps away from a landscape's most beautiful feature.

He began.

"Yes, that lake bottom soil was so rich, we'd get corn twelve maybe fifteen feet high in the dry years. Then in the forties the water came back. The low land on the farm needed good drainage into the lake to raise crops. Dad eventually put in drain tile. Of course, the livestock went into the lake to drink and cool themselves until about twenty years ago, when they started aborting their calves. The vet said it was because they were drinking something bad from the lake.

"Years ago a man from the Park District stopped by and told Dad they'd like to buy some of our land that touched the lake as part of a 3,500-acre park they were setting up. Dad simply said, 'We're dairying here,' and sent him on his way."

Henry, neither my farmers nor yours saw lakescape as nature untrammeled, as pastoral scene to be appreciated for its aesthetic qualities. They saw the lake as but a part of the mosaic of human habitat viewed from a different rung on the ladder of human needs, a part of the garden, part of the place humans make home.

I must admit I now cringe at your shrill characterization of Farmer Flint and others that "see" lakescape as neither you nor I have. I now see the paradox I set out to understand results not so much from failure of perception, as I first expected, though I think that failure is real and common and injurious to lakes in ways all would decry. Rather the paradox arises out of an amalgam of differing, sometimes conflicting visions of the same lakescape, where the "hard bottom of truth" lies more deeply buried than either of us expected.

Henry, your diatribe against Farmer Flint declares that he neither knew nor loved the lake, and he exhausted the land and would have exhausted the water, too, if he had been able. I've heard that decades after you passed on, a thousand shallow lakes in the 3,000-acre-size class were drained to make farmland in Minnesota, converting lakescape perceived as "problem" into landscape perceived as habitat for humankind.

In your day the townspeople of Lincoln saw it right that farmers raise crops in this lakescape, saw this lakescape as human habitat. The towns-folk today see it differently, see it as a drinking water source, yes, but also something much more. Restricting use of road salt, chasing geese and their phosphorus and bacteria-laden poop away, prohibiting swimmers and their pee, the lawn fertilizer ban, and the walking prohibition close to the water—the town sees this pond not as an isolated pool of drinking water, but water inextricably linked with a landscape, part of an interconnected land-lake system. I find that exciting and very significant.

I am struck, Henry, that regardless of our differing visions of lakes, no lake is enhanced by the runaway ills that now plague them. Aside from those who see shallow lakes as problems to be drained, all other visions of lakes ultimately share the same goals, goals reachable only through enlightened stewardship. That's what excites me about the symbolism of Flint's Pond.

Henry, I think of your statement in *Walden* that "a lake is the landscape's most beautiful and expressive feature. It is earth's eye; looking into which the beholder measures the depth of his own nature." Those words today are the totem of lake protection activists across the country. At the outset of my journey, your words struck me as more clever than clear. How does one measure the "depth of his own nature" by looking into a lake? Your words make sense to me now. Looking into earth's eye reflects our values, beliefs, and behaviors.

A lake is an archive of all that we do. What we do to the land and to the air we do to the lake. Lake as earth's eye is more than metaphor.

Darkhouse

*Our perceptions of nature and how it works often tell us less about
what is actually out there in the landscape and more about the
types of mental topography, biases and projections that we carry
about with us in our heads.*

DUNCAN TAYLOR

Cold snow crunches beneath our boots as Geri and I make our way across
the frozen surface of Rainy Lake toward a small dark shack, one of several
fish houses scattered across the glistening white surface. Unlike fish houses
of anglers, this one has no windows. This is a darkhouse where fish are
taken by spear.

Darkhouse spearing of northern pike has been a long tradition in my
family. Uncle Harold introduced my dad and me to spearing when I was
nine. From grade school days through college winter breaks to my first years
of married life, Dad and I spent hours in the darkhouse together.

I open the door and step in. It feels like home. A hole in the ice, three
feet long and three feet wide, dominates the closetlike space. An inch-thick
layer of ice has formed over the hole since the house was last used. I break
the ice skim with a hatchet, strain the floating ice chips from the water and
fling them out the door. The kerosene stove lit, we take our seats on a narrow bench along a wall, the toes of our boots at the edge of the hole. I pull
the door shut, and we are bathed in darkness. Eyes soon accommodate to
light from the hole, and I unwind the fish line from my decoy, tie one end to

a tiny screw-eye on its back and the other to a nail in the wall across from the hole, and lower into the lake the gray-and-white wooden fish with tin fins, thumbtack eyes, and a small slug of lead in its belly. A bend in the tail fin causes the fish to descend into the darkened water in a lazy spiral until it reaches the end of its tether in the center of the hole, maybe six feet down.

Black paint covers the ceiling and walls to ensure that no light or reflections pass from the house into the hole, alerting fish to our presence. Its rope attached to the shack's wall, I lower the head of the spear into the water, hooking a tine into the ice. A jerk on the decoy's line and the imposter fish glides through a circle and a half before it returns to rest. We stare into amorphous green-tinted nothingness and wait.

I cannot see bottom. The mind cannot abide bottomlessness. We need reassurance. I remove a fresh potato from my pocket, cut it into thin slices and drop them into the water. The discs descend in a back and forth rocking motion and settle scattered across the bottom. Sunlight penetrating the snow and ice reflects off the white patches. Bottom. Perspective established.

· · ·

I do not understand myself. I am an impatient fisherman with hook and line. If no fish bite within half an hour, I'm anxious to stash the rod and move on to other things. Not so in a darkhouse. I can peer into a spearing hole all day.

Inside the shack the world is simple and dark, warm and quiet. Ostensibly, a darkhouse is about taking fish. For Dad and me it became something more. Except for the gray and white of the decoy, the occasional twang of the plucked tether putting the fake fish in motion, and the faint odor of kerosene, the eyes and ears and nostrils have little to report. No longer burdened by the cacophony of sensory messages that usually beset the mind, it opens itself to thought. Minds engage in low, near-whispered conversation separated by long intervals of silence. Conversations about feelings, aspirations, life decisions, and world affairs punctuate the silence. More than the taking of fish, I grew to deeply value those quiet contemplations with Dad.

· · ·

My formal exploration of the love-lakes-but-degrade-them paradox ends here, on Rainy Lake, where it began. Understanding the paradox that sparked this journey turned out to be far more complex than I had expected. Geri and I have come to this darkened womb to reflect, to distill the insights from the journey, to give birth to understanding.

The stove has finally overcome the cold, and we remove bulky coats, hang them on nails, then peer back into our window on the lake. "When I was a kid," I tell Geri in a hushed voice, "Dad was scared stiff I'd fall into the hole. He finally rid himself of anxiety by attaching me by a rope to the wall."

"I'll pull you out if you fall in," Geri whispers back.

We notice the straight shaft of the spear appears to bend at the point it enters the water. "That's what happens when an object passes from a medium of one optical density into another," Geri says, ever the science teacher.

"It also confuses perception," I add. I think about how my Uncle Harold let me, as a kid, use a darkhouse he had put out on a small lake just blocks from his work. During his lunch break he'd trudge across the ice to see how I was doing. The sound of boots grew nearer, then finally stopped outside the door, and he would ask through the house wall how many pike I had seen. The truth, almost always, was that I had seen nearly a dozen. Ashamed, I had to admit that I had missed every one. I ultimately learned that the chance of hitting a fish depends directly on whether I throw the spear on the angle of the shaft above water or the angle below. Perception matters.

· · ·

A pike! Far side of hole. Running deep. Grab the spear . . . Gone . . . Disappeared into the nebulous green-brown. "Nice fish," Geri exclaims.

"He never slowed up."

I replace the spear's tine into its pocket in the ice. Normality returns below, the adrenalin rush subsides above and minds settle back into thought.

Geri finally breaks a long silence. "Are you still struggling with the humans and lakes contradiction?" she asks.

After a lengthy pause, I reply. "I think people truly do love lakes, often deeply, but the mind has innate tendencies, comfortable habits, that wedge themselves between our love and our lakes. Mental tendencies appear often to nudge reason aside and shape perceptions in ways that produce behaviors with unintended consequences."

Our instinctive attraction to the savanna-by-the-quiet-water illustrates what I mean. Intellectually, I understand the impact on Dream Lake were we to remove much of the vegetation and convert our lot mostly to lawn. But, oh, how powerfully my savanna instinct urges me to do so. My mind craves an opened landscape.

"So you feel that the ancient emotional pull of the savanna mindscape blinds us to the degradation of the lake that results?"

"Yes," I reply.

"But that conflicts with what most of us believe about ourselves."

Other tendencies influence our perceptions. Just as a home town crowd at a basketball game sees floor actions differently than the referee, we who live in lake watersheds are inclined to see our activities as more benign than people with hard facts and data do. We can so easily see the world the way we wish it to be instead of the way it is.

Unfortunately, the nature of lake degradation itself can confound our perceptions. It would have been difficult, I expect, to make the argument in the 1960s that Milltown's Rice Lake was headed for trouble. Reason might have foreseen that, given the substantial nutrient loading into the lake during earlier decades, a time of consequence would ultimately come. But imagine the skeptic's argument: events on the land had gone on for decades with little noticeable effect, so what's the basis for worry? Our tendency to focus on the here and now has largely served our species well, but it has compromised our ability to glimpse the future.

Perhaps our biggest obstacle to transforming our love of lakes into informed stewardship has been our tendency to see the world as collections of objects and categories of classification. That habit of mind long held back even the hard-nosed objectivity of science from seeing lake as a system. Failure to understand a lake as a system is failure to understand a lake's nature. Absent understanding, our love of lakes must ever remain unconsummated.

. . .

"Look," Geri whispers loudly, "a fish!" A pike moving slowly eyes the resting decoy from behind.

I equivocate. Should I throw? Back in college days, when two twelve-pound pike came through the hole forty-five minutes apart, I nailed them both. Then I closed up the shack for the day and left for home, my mind flitting between the exuberance at such undreamed of success and contrition at the ease and speed with which it had been accomplished. I did not return to the spear house the rest of that Christmas vacation. I have never enjoyed the act of killing with a spear, though I've done it many times. But if fish are to be eaten, they must first be taken from the lake.

My dilemma today is more complicated. Fishes face new challenges in this age of lake degradation. While much of Rainy Lake remains wild and is not beset by the kinds of problems facing many lakes elsewhere, the symbolism of going after fish with a spear discomfits me. Thoreau seems to have had no such reluctance taking fish. Though he went often onto Walden in winter, I don't know that he ever speared. He did admit to discovering pike resting in shallows in other seasons and whacking one with great force, writing, "I cut him almost in half."

The pike decides to leave the decoy alone and moves to the edge of the hole before an ancient instinct kicks in, and I grab the spear and throw. Bad angle. I know I have missed him even as the spear hurtles through the water, turning the hole into a bubble chamber. The rope has checked the racing spear and leaves it dangling a foot above the lake bottom. I pull it up and return a tine to its pocket in the ice.

·   ·   ·

Excitement ebbs and Geri returns me to thought. "So you see a conflict between emotions, including those tendencies of the mind, as you call them, and objective reason."

"I do," I reply. Two neuroeconomists, Colin Cameron and George Loewenstein, sum my thoughts well: "The mind is a charioteer driving twin horses of reason and emotion. Except cognition is a smart pony, and emotion is an elephant."

"So you see the paradox as the outcome of a struggle between reason and emotion, where emotion too often wins," Geri's voice trails off.

I see by the look on her face, illumined by light flowing up from the hole, the science teacher is unsure that emotion could outweigh reason. "Fistfuls of evidence suggest that's the way we are," I say, then describe the findings of public health expert Valerie Curtis. More than a million lives are lost worldwide each year because people simply fail to wash their hands, an astounding figure given that the importance of hand-washing has been one of the most widely disseminated health messages, its powerful effectiveness undisputed. Education efforts have failed miserably in developed and undeveloped countries alike. Curtis reports on studies showing people are much more likely to wash their hands if they are motivated to do so at a deep emotional level. Mothers shown a video of a caring mother accidentally contaminating her child's food after having gone to the toilet became six times more likely to wash their hands than other mothers exposed only to factual education.

Curtis concludes, "To break unhealthy habits, campaigns need to target the emotions, because they are the decision makers. Where the heart leads, the habits will follow."

• • •

Our attention returns to the hole, and I jerk the decoy string. Bang—in a flash—a pike roars in from the left and slams the decoy. He holds it for a moment in his mouth, then lets go, backs away, then stops. My emotions divide. One inner voice sympathizes. Let him be, it says. *Let him be.* The hunter's instinct says food, *take food.* Reason says look here, this species is thriving. His habitat is not in danger. Another species, other circumstances, maybe hold back. *But not here, not now.* Instinctively I finger the spear, lower it silently into the water, and throw as he turns to leave, aiming ahead of the fish to compensate for his movement. Pandemonium breaks out below—bubbles, rope, steel shaft, twisting decoy. I pull the spear up by its rope through a shower of scales, like spangles, carnival glitter, that settle slowly to the lake floor. The spear is heavy with fish. I open the door into blinding light and shake the fish off onto the snow. "Nice sized," I call back to Geri. "Four, maybe five pounds." Enough for several meals. A fish this size most certainly has mercury in its flesh. We will freeze it at home and space meals of fish monthly.

I follow the time-honored tradition of the spearing fraternity and gather every speck of bloodied snow and ice and bury them alongside the fish in a snowbank. No point attracting other spearers and their houses.

Exhilarated, we take advantage of the opened door and walk vigorously to stretch cramped legs and backs before returning to our seats in the shack to resume our vigilance. Adrenalin continues to rev our hunter brains and slowly, ever so slowly, our minds resume thoughts of lake and paradox.

. . .

I find myself conflicted. The scientist in me roots for reason. My brain wants to dump those habits of mind, to see lakes more objectively. But it is emotion that draws me to lakes.

"So, what about Thoreau?" Geri interjects.

"Thoreau is a conundrum," I reply. He says he wants those emotional habits of mind to romp freely, to shove aside science and reason. He scorns science at nearly every turn, yet often embraces it. His romantic vision of nature is given credit for giving birth to environmental concerns, yet in a hundred fifty years that vision has not adequately defended those precious "eyes of earth" he so loved. In that sense Thoreau's romanticism failed; yet, paradoxically, it nurtures concern for nature to this day. Romanticism provides the love, but reason provides a check on reality.

I think about how our bias to see lakes as a collection of objects has hindered our reaching the hard bottom of truth that lakes are functional and interacting systems that include land and air. If we had understood that much earlier, I like to think it would have blunted our degradation of lakes. Comparable recognition of systems likely would have hastened our "seeing" the reality of global warming.

I think about the pull of science, of objective reasoning, that Thoreau denied but ultimately found impossible to resist. The most scientifically significant of Thoreau's essays, "The Succession of Forest Trees," was published less than two years before his death. It is a masterpiece of objective, reasoned thinking, now generally acknowledged as breaking first ground in a new science we today call ecology. I have read his essay carefully many times, searching for clues that Thoreau may have moved beyond the purely descriptive side of nature observation and stuck a toe into the world of nature seen as function and system. I think he arrived right at the edge.

Might our lakes have fared better had Thoreau's thinking moved still deeper into science and, instead of dying at forty-four years of age, he had been given forty years more? Plankton and waterfleas were recognized not long after his death. With Thoreau's curiosity they would have been as irresistible to him as they were to Birge and Forbes. Might such a fascination have led him to see lake as community sooner than Forbes?

What an important bridge Thoreau might have become between Arcadian romanticism and the hard numbers of science. Emotion and reason combined. Joseph Wood Krutch implies as much when he writes of Thoreau: "To unite without incongruity things ordinarily thought of as incongruous *is* the phenomenon called Thoreau." While to suggest that Thoreau might have had a major impact on our understanding of lakes is speculation, his intense curiosity and acute powers of observation convince me the idea cannot be rejected out of hand.

Thoreau leaves little doubt about how he perceived the pursuit of truth: "Let us settle ourselves and work and wedge our feet downward through the mud and slush of opinion, and prejudice, and tradition, and delusion, and appearance . . . through church and state, through poetry and philosophy and religion, till we come to a hard bottom and rocks in place which we can call reality, and *say* This is, *and then begin.*"

I cannot help but wistfully wonder what might have been.

. . .

Though I'm staring down the hole as I think, I have all but forgotten pike. The vague form of a fish appears motionless down and right of the decoy. I am unsure. Geri says she sees nothing. In minutes my eyes, or is it my mind, sees another shadowy fish form. Again it does not move. Again, Geri sees nothing. Dad and I occasionally saw such apparitions, most often when heavy clouds or low angle of the winter sun darkened the hole.

Light now dimming, it is time to leave. I retrieve the decoy and notice new teeth marks in the wood, battle scars of the day. I open the door and dig our fish out of the snowbank as Geri loads decoy, empty lunch bags, water bottles, and the spear onto the sled.

. . .

I reflect in the car as we head for home. The continental glacier delivered a most precious gift. As ancestors of our tribe busied themselves inventing agriculture merely 10,000 years ago, the glaciers were putting finishing touches on the lakes. Yet how ephemeral are those creations. Already, the once rich lake district of the American Southwest is gone. Only a smattering of lakes survive there, testimony to that lake district's former existence, a victim of climate change 12,000 years ago. Lake sediment cores across the continent reveal that since the glaciers left, most of the original depth of many lake basins has already been filled in with mud and sand. Many of our lakes approach the autumn of their lives speeded along by our actions. Mountains may last for eons of geologic time, but lakes for only a geological moment. It is our cosmic good fortune to be on earth at such a moment.

I think of our time in New Hampshire on Mirror Lake. Likens and Bormann did more than confirm the atmosphere as a significant part of lake systems. Their studies implicitly reveal we humans are a more diverse and significant presence in those systems than we had imagined. But unlike the chemical, physical, biological, geological, and atmospheric overlays of the lake system, the human overlay has the power to perceive, to see, to understand, and to act.

Eutrophication, erosion, pollution, exotic species invasions, and removal of shore vegetation and native aquatic plant communities accelerate lakes through their already too short natural lives. Inaction does not become creatures of love, common sense, and reason. Descendants of our tribe would never forgive us.

Absent love, lake protection is an impossible mission. Absent understanding, lake protection is a fool's dream.

What will be our tribal narrative, Henry? What will be our legacy?

SELECTED BIBLIOGRAPHY

Abram, David. *The Spell of the Sensuous: Perception and Language in a More-than-Human World.* New York: Pantheon Books, 1996.

Ackerman, Diane. *A Natural History of the Senses.* New York: Vintage Books, 1990.

Bascom, Willard. *The Dynamics of the Ocean Surface.* Rev. and updated ed. Garden City, NY: Anchor Books/Doubleday, 1980.

Bloomfield, J. A. *Lakes of New York State: Ecology of the Finger Lakes,* 1. New York: Academic Press, 1978.

Bolles, Edmund B. *The Ice Finders: How a Poet, a Professor, and a Politician Discovered the Ice Age.* Washington, DC: Counter Point, 1999.

Borman, Susan, Robert Korth, and Jo Temte. *Through the Looking Glass: A Field Guide to Aquatic Plants.* Stevens Point: Wisconsin Lakes Partnership and University of Wisconsin–Stevens Point, 1999.

Broadhurst, C. Leigh, Stephen C. Cunnane, and Michael A. Crawford. "Rift Valley Lake Fish and Shellfish Provided Brain-Specific Nutrition for Early Homo." http://journals.cambridge.org/action/displayJournal?jid=BJN. *British Journal of Nutrition* 79 (1998): 3–21.

Brönmark, Christer, and Lars-Anders Hansson. *The Biology of Lakes and Ponds.* New York: Oxford University Press, 1998.

Buell, Lawrence, ed. *Thoreau's Sense of Place: Essays in American Environmental Writing.* Iowa City: University of Iowa Press, 2000.

Burgis, Mary, and Pat Morris. *The Natural History of Lakes.* New York: Cambridge University Press, 1987.

Carlson, Bruce M. *Beneath the Surface: A Natural History of a Fisherman's Lake.* St. Paul: Minnesota Historical Society Press, 2007.

Chorus, Ingrid. "Toxic Cyanobacteria: Controlling the Risk." *Lakeline* 26, no. 2 (Summer 2006): 16–23.

Climate Change Connection. "Climate Change and Lake Winnipeg." Http://www.climatechangeconnection.org/impacts/LakeWinnipeg.htm.

Cole, Gerald A. *Textbook of Limnology.* 3rd ed. St Louis: Mosby Co., 1983.

Correll, David L. "The Role of Phosphorus in the Eutrophication of Receiving Waters: A Review." *Journal of Environmental Quality* 27 (1998): 261–66.

Deevey, Edward S., Jr. "A Re-examination of Thoreau's 'Walden.'" *Quarterly Review of Biology* 17, no. 1 (March 1942): 1–11.

Dillard, Annie. *Pilgrim at Tinker Creek.* New York: Harper & Row, 1974.

Dudiak, Tamara, and Robert Korth. *How's the Water? Planning for Recreational Use on Wisconsin Lakes and Rivers.* Stevens Point: Wisconsin Lakes Partnership, 2002.

Dupre, A. Hunter. "Thoreau as Scientist: American Science in the 1850s." In *Thoreau's World and Ours: A Natural Legacy,* edited by Edmund A. Scholfield and Robert C. Baron, 42–47. Golden, CO: North American Press, 1993.

Engel, Sandy, and Stanley A. Nichols. "Aquatic Macrophyte Growth in a Turbid Windswept Lake." *Journal of Freshwater Ecology* 9, no. 2 (June 1994): 97–109.

———. "Restoring Rice Lake at Milltown, Wisconsin." *Wisconsin Department of Natural Resources, Technical Bulletin,* no. 186 (1994).

Fink, David. *A Guide to Aquatic Plants: Identification and Management.* St Paul: State of Minnesota, Department of Natural Resources, 2000.

Forbes, Stephen A. "The Lake as a Microcosm." In *Foundations of Ecology: Classic Papers with Commentaries,* edited by Leslie A. Real and James H. Brown, 14–27. Chicago: University of Chicago Press, 1991.

Frey, David G., ed. *Limnology in North America.* Madison: University of Wisconsin Press, 1963.

Gallagher, Winifred. *The Power of the Place: How Our Surroundings Shape Our Thoughts, Emotions, and Actions.* New York: Harper Perennial, 1993.

Gibbons, Ann. *The First Human: The Race to Discover Our Earliest Ancestors.* New York: Doubleday, 2006.

Goggin, Patrick O., Daniel Haskell, and Michael M. Meyer. "Wisconsin Lakeshore Restoration." *Lakeline* 29, no. 1 (2009): 23–27.

Graham, Jennifer L., John R. Jones, and Susan B. Jones. "Microcystin in Midwestern Lakes," *Lakeline* 26, no. 2 (2006): 32–35.

Gregory, Stanley V., Frederick J. Swanson, W. Arthur McKee, and Kenneth W. Cummins. "An Ecosystem Perspective of Riparian Zones." *Bioscience* 41 (1991): 540–51.

Gunn, Jim, R. J. Steedman, and R. A. Ryder, eds. *Boreal Shield Watersheds: Lake Trout Systems in a Changing Environment.* Boca Raton, FL: Lewis Publishers-CRC Press, 2004.

Hagen, Joel B. *An Entangled Bank: The Origins of Ecosystem Ecology.* New Brunswick, NJ: Rutgers University Press, 1992.

Halpern, Daniel, ed. *On Nature: Nature, Landscape, and Natural History.* San Francisco: North Point Press, 1987.

Henderson, Carrol L., Carolyn J. Dindorf, and Fred J. Rozumalski. *Landscaping for Wildlife and Water Quality.* St. Paul: Minnesota Department of Natural Resources, Non Game Wildlife Program, n.d.

Heyes, A., T. R. Moore, J. W. M. Rudd, and J. J. Dugova. "Methyl Mercury in Pristine and Impounded Boreal Peatlands, Experimental Lakes Area, Ontario." *Canadian Journal of Fisheries and Aquatic Sciences* 57, no. 11 (November 2000): 2211–22.

Hoff, Mary. "New Visions for Lake Shores." *Minnesota Conservation Volunteer* 68, no. 401 (July–August 2005): 9–19.

Horne, Alexander J., and Charles R. Goldman. *Limnology.* 2nd ed. St. Louis: McGraw-Hill, 1994.

Hurd, Barbara. *Stirring the Mud: On Swamps, Bogs and Human Imagination.* Boston: Beacon Press, 2001.

Kallemeyn, Larry W., Kerry L. Helmberg, Jim Perry, and Beth Y. Odde. "Aquatic Synthesis for Voyageurs National Park." *U.S. Geological Survey, Biological Resources Division Information and Technology Report 2003–0001* (May 2003).

Kay, Jane Holtz. "Wall to Wall at Walden." *Nation* 246, no. 24 (June 18, 1988): 867–72.

Kellert, Stephen, and Edward O. Wilson, eds. *The Biophilia Hypothesis.* Washington, DC: Island Press, 1993.

Lakoff, George, and Mark Johnson. *Metaphors We Live By.* Chicago: Chicago University Press, 1980.

Laland, Kevin N., and Gillian R. Brown. *Sense and Nonsense: Evolutionary Perspectives on Human Behavior.* New York: Oxford University Press, 2002.

Lampert, Winfried, and Ulrich Sommer. *Limnoecology: The Ecology of Lakes and Streams.* Translated by James F. Haney. New York: Oxford University Press, 1997.

Leakey, Richard, and Roger Lewin. *People of the Lake: Mankind in Its Beginnings.* New York:

Doubleday, 1978.

Leopold, Aldo. *A Sand County Almanac: And Sketches Here and There.* New York: Oxford University Press, 1949.

Lewis, Pierce F. "Axioms for Reading the Landscape: Some Guides to the American Scene." In *The Interpretation of Ordinary Landscapes: Geographical Essays,* edited by D. W. Meinig, 11–32. New York: Oxford University Press, 1979.

Likens, Gene E., ed. *An Ecosystem Approach to Aquatic Ecology: Mirror Lake and Its Environment.* New York: Springer-Verlag, 1985.

Lindeman, Ray L. "The Trophic-Dynamic Aspect of Ecology." *Ecology* 23, no. 4 (1942): 399–417.

Maynard, W. Barksdale. *Walden Pond: A History.* New York: Oxford University Press, 2004.

McKnight, Diane M. "Regional Assessment of Freshwater Ecosystems and Climate Change in North America." *Symposium Report.* Leesburg, VA, October 24–26, 1994.

Meine, Curt. *Correction Lines: Essays on Land, Leopold, and Conservation.* Washington, DC: Island Press, 2004.

Meinig, D. W. "The Beholding Eye: Ten Versions of the Same Scene." *Landscape Architecture* (January 1976): 47–54.

———. "Introduction." In *The Interpretation of Ordinary Landscapes: Geographical Essays,* edited by D. W. Meinig, 1. New York: Oxford University Press, 1979.

Merritt, Richard W., and Kenneth W. Cummins, eds. *An Introduction to the Aquatic Insects of North America.* 3rd ed. Dubuque: Kendall-Hunt, 1996.

Meyers, Steven J. *On Seeing Nature.* Golden, CO: Fulcrum, 1987.

Mortimer, C. H. *An Explorer of Lakes.* Madison: University of Wisconsin Press, 1956.

Naiman, Robert, J. J. Magnuson, D. M. McKnight, and J. A. Stanford. *The Freshwater Imperative: A Research Agenda.* Washington, DC: Island Press, 1995.

Nelson, Darby. "Canaries of Deep Water." *Minnesota Conservation Volunteer* 71, no. 419 (July–August 2008): 20–27.

———. "Waterflea (Genus Daphnia)." *Minnesota Conservation Volunteer* 68, no. 401 (July–August 2005): 64.

Nelson, Richard K. *Make Prayers to the Raven: A Koyukon View of the Northern Forest.* Chicago: University of Chicago Press, 1983.

Nesbitt, Richard E. *The Geography of Thought: How Asians and Westerners Think Differently.* New York: New York Free Press, 2003.

Norvald Fimreite, H. G. "Mercury Contamination of Aquatic Birds in Northwestern Ontario." *Journal of Wildlife Management* 38, no. 1 (1974): 120–31.

Ojakangas, Richard W., and Charles Matsch. *Minnesota's Geology.* Minneapolis: University of Minnesota Press, 1982.

Olson, Sigurd F. *Open Horizons.* New York: Alfred A. Knopf, 1969.

Orians, Gordon H., and Judith H. Heerwagen. "Evolved Responses to Landscapes." In *The Adapted Mind: Evolutionary Psychology and the Generation of Culture,* edited by Jerome Barkow, Leda Cosmides, and John Tooby, 555–80. New York: Oxford University Press, 1992.

Pennak, Robert L. *Freshwater Invertebrates of the United States.* New York: Ronald Press, 1953.

Perkins, Sid. "Once upon a Lake: The Life, Times, and Demise of the World's Largest Lake." *Science News* 162 (2002): 283–84.

Phillips, Gary, William Schmid, and James C. Underhill. *Fishes of the Minnesota Region.* Minneapolis: University of Minnesota Press, 1982.

Phillips, Nancy, Martin Kelly, Judith Taggart, and Rachel Reeder. *The Lake Pocket Book.* Alexandria, VA: Terrene Institute with U.S. Environmental Protection Agency, Region 5, 2000.

Pielou, E. C. *After the Ice Age: The Return of Life to Glaciated North America.* Chicago: University of Chicago Press, 1991.

Power, Rebecca, Ken Genskow, and Linda Prokopy. "Exploring the Social Dimensions." *Lakeline* 28, no. 3 (Fall 2008): 15–18.

Price, Jeremy. "Better Lakes for Tomorrow." *Lakeline* 29, no. 1 (2009): 32–36.

Radomski, Paul, and Timothy J. Goeman. "Consequences of Human Lakeshore Development on Emergent and Floating-Leaf Vegetation Abundance." *North American Journal of Fisheries Management* 21 (2001): 46–61.

Raloff, Janet. "Why the Mercury Falls: Heavy Metal Rains May Trace to Oxidants, Including Smog." *Science News* 163 (2003): 72–74.

Rossi, William. "Thoreau as a Philosophical Naturalist-Writer." In *Thoreau's World and Ours: A Natural Legacy,* edited by Edmund A. Scholfield and Robert C. Baron, 64–73. Golden, CO: North American Press, 1993.

Russell, Francis. *Mistehay Sakahegan—The Great Lake: The Beauty and Treachery of Lake Winnipeg.* Winnipeg: Heartland Publications, 2000.

Sanders, Scott Russell. *A Conservationist's Manifesto.* Bloomington: Indiana University Press, 2009.

———. *Hunting for Hope: A Father's Journeys.* Boston: Beacon Press, 1998.

———. *Staying Put: Making a Home in a Restless World.* Boston: Beacon Press, 1993.

Sattlemeyer, Robert. "The Coleridge Influence on Thoreau's Science." In *Thoreau's World and*

Ours: A Natural Legacy, edited by Edmund A. Scholfield and Robert C. Baron, 48–54. Golden, CO: North American Press, 1993.

Schama, Simon. *Landscape and Memory.* New York: Vintage Books, 1995.

Scheffer, Marten. *Ecology of Shallow Lakes.* New York: Chapman & Hall, 1998.

Scholfield, Edmund A., and Robert C. Baron, eds. *Thoreau's World and Ours: A Natural Legacy.* Golden, CO: North American Press, 1993.

Sellery, G. C. "E. A. Birge: A Memoir: With an Appraisal of Birge the Limnologist." In *An Explorer of Lakes,* edited by C. H. Mortimer, 165–211. Madison: University of Wisconsin Press, 1956.

Sewall, Laura. *Sight and Sensibility: The Ecopsychology of Perception.* New York: Jeremy P. Tarcher/Putnam Penguin, 1999.

Thoreau, Henry David. *The Essays of Henry David Thoreau: Selected and Edited by Lewis Hyde.* New York: North Point Press, 2002.

———. *The Portable Thoreau: Walden, Essays, Poems, Journal.* New York: Viking Press, 1964.

———. *Thoreau on Water: Reflecting Heaven.* Edited by Robert Lawrence France. New York: Houghton Mifflin Co., 2001.

Thorp, James H., and Alan P. Covich, eds. *Ecology and Classification of North American Freshwater Invertebrates.* San Diego: Academic Press, 1991.

Thorson, Robert M. *Beyond Walden: The Hidden History of America's Kettle Lakes and Ponds.* New York: Walker and Co., 2009.

Tuan, Yi-Fu. *Topophilia.* New York: Columbia University Press, 1990.

U.S. Environmental Protection Agency. *National Lakes Assessment.* Http://www.epa.gov/owow/lakes/lakessurvey/.

U.S. Geological Survey, U.S. Department of the Interior. "Walden Pond, Massachusetts: Environmental Setting and Current Investigations." *USGS Fact Sheet FS-064–98* (June 1998). Http://ma.water.usgs.gov/publications/pdf/wal_66.pdf .

———. "Water Quality of Lakes and Streams in Voyageurs National Park, Northern Minnesota 1977–84." *Water Resources Investigation Report* 88–4016.

Vennum, Thomas, Jr. *Wild Rice and the Ojibway People.* St. Paul: Minnesota Historical Society Press, 1988.

Walls, Laura Dassow. "Seeing New Worlds: Thoreau and Humboldtian Science." In *Thoreau's World and Ours: A Natural Legacy,* edited by Edmund A. Scholfield and Robert C. Baron, 55–63. Golden, CO: North American Press, 1993.

Westfall, Minter J., Jr. *Damselflies of North America.* Gainesville, FL: Scientific Publishers, 1996.

Wetzel, Robert G. *Limnology: Lake and River Systems.* 3rd ed. San Diego: Academic Press, 2001.

Willoughby, Pamela R. *The Evolution of Modern Humans in Africa: A Comprehensive Guide.* Lanham, MD: Rowman & Littlefield, 2007.

Worster, Donald. *Nature's Economy: A History of Ecological Ideas.* 2nd ed. New York: Cambridge University Press, 1994.

Zumberge, James. "The Lakes of Minnesota: Their Origin and Classification." *Minnesota Geological Survey Bulletin* 35. Minneapolis: University of Minnesota Press, 1952.